Environmental Microbiology

Alan H. Varnam

BSc, PhD
Consultant Microbiologist
Southern Biological and
Visiting Lecturer
University of North London, UK

Malcolm G. Evans

Photographer

ASM
PRESS

WASHINGTON, D.C.

First published in the United States of America in 2000 by:
ASM Press,
1752 N Street, NW,
Washington, DC 20036–2804.

ISBN 1–55581–218–X

Library of Congress Cataloging-in-Publication Data applied for.

Project management, design and layout: Paul Bennett
Cover design: Patrick Daly
Text editing: Michael Meakin
Color reproduction: Tenon & Polert Colour Scanning Ltd, Hong Kong
Printed by: Grafos SA, Barcelona, Spain

CONTENTS

PREFACE

Adam had 'em
The world's shortest poem: on the antiquity of microbes

Micro-organisms are ubiquitous. They are present in the soil of our gardens, in the air that we breathe and in the sea around us. They invade and colonize our bodies, both causing illness and being protective of health. They spoil our foods and yet are involved in the manufacture of some. Micro-organisms are able to grow in a remarkable range of habitats, which range from the relatively hospitable, often supporting a wide diversity of microbial life, to the extremely hostile, where only a single type may proliferate. An example of an extremely hostile environment, which has attracted much recent interest, is the vent of undersea volcanoes – the realm of the hyper-thermophile. Despite enormous differences in the nature and geobiology of the various environments, it should be appreciated that the same underlying factors govern the life and death of micro-organisms irrespective of the nature of the environment. Equally, their behaviour is governed by fundamental rules.

A vast amount of work, both at bench and theoretical level has been necessary over a period of many years to enable us to assemble the body of knowledge concerning environmental microbiology available today. Until relatively recently, bench work has been primarily based on conventional cultural techniques and, providing the inherent limitations are understood, these can still be of considerable value. In recent years, however, an array of powerful new tools, many based on genetic or advanced microscopic techniques, have become available. Advances at bench level have been complemented by the development of new concepts at the theoretical level.

The study of environmental micro-organisms is not merely an academic pursuit. Environmental micro-organisms have long been exploited, knowingly or unknowingly, in the management of Planet Earth and in the economic activities of humans. Our increased understanding of the interactions between micro-organisms and the environment have enabled a refinement of their exploitation. An example is the bioremediation of soil polluted by oil during the Gulf War.

This book is intended to provide a sound, fundamental knowledge of environmental micro-biology, the principles being reinforced by a wide range of examples, illustrated in colour where appropriate. The interactions between the environment and the dominant microflora is discussed as are the interactions between the various components of the microflora. The emphasis, however, is on understanding principles. This book is neither intended to be a long list of every micro-organism which may be present in a given environment, nor is it intended to provide detailed descriptions of the taxonomy of the micro-organisms. Such information, if required, may be obtained elsewhere.

As environmental microbiology is a difficult subject area to divide into well defined and rigid categories, the usual chapter-based format has not been used. The book is divided into four main parts: an Overview of environmental microbiology, Aquatic, Terrestrial and Extreme environments, each subdivided into major subject areas. It is intended that this rather unconventional approach will enhance the ease of use. Above all, however, it is hoped that the book will go at least some way to show what a challenging and fascinating topic environmental microbiology is, even when experimental results appear to defy interpretation.

Alan H. Varnam
Malcolm G. Evans

GLOSSARY

Definitions are given in the context of micro-organisms, although similar terms may be applied to other forms of life.

Acidophile. A micro-organism which grows optimally at pH values below 5.0. Should not be confused with acid-tolerant micro-organisms. A **thermoacidophile** grows both at high temperatures and low pH value.

Actinomycete. A diverse group of Gram-positive eubacteria, amongst which a mycelial growth habit is common.

Alkaliphile. A micro-organism which grows optimally at pH values above 8.0.

Allochthonous. Micro-organisms which are not consistently a major part of the microflora of an environment, but which become of significance in response to an increase in nutrient concentrations, or other favourable event. Known previously as **zymogenous**.

Archaebacteria. A primary subgroup of cellular life, comprising bacteria of prokaryotic cell structure, but a highly distinctive cell chemistry, which differs profoundly from that of eubacteria. Also known collectively as **archae** and individually as an **archeon**.

Autochthonous. Micro-organisms which are consistently present in significant numbers in an environment and form the indigenous microflora.

Autotroph. An organism capable of growth using only inorganic nutrients. Typified by green plants, autotrophic micro-organisms include **photoautotrophs** (light as their energy source and CO_2 as their C source) and **chemoautotrophs** (chemicals as their energy source). Also known as **chemolithotrophs**.

Bloom. Seasonal dense growth of algae or other planktonic micro-organisms on water.

Copiotroph. A micro-organism which requires a high level of nutrients for growth.

Coryneform. Gram-positive bacteria, of characteristic irregular morphology, similar to that of *Corynebacterium*. These are numerically important in a wide range of habitats, including soil.

Cyanobacteria. Photosynthetic eubacteria, previously known as 'blue-green' algae.

Denitrification. Anaerobic bacterial respiration of nitrate, resulting in reduction to nitrite and, by a cascade of two further anaerobic respirations, to N_2 gas. As part of the nitrogen cycle, it is important in geomicrobiology, but also occurs in nitrite-containing (cured) foods.

Epiphyte. A micro-organism which grows on the surface of plants, but takes no water or nutrients from the plant.

Eubacteria. A primary subgroup of cellular life, comprising most bacteria. Cells are prokaryotic in structure, but the cell chemistry is similar to that of eukaryotes.

Eukaryote. An organism in which the unit of structure is the complex eukaryotic cell. Eukarotes are of extraordinary diversity, including all known life forms except the prokaryote*s*.

Eutrophication. The process of enrichment of a body of water (usually freshwater) with inorganic nutrients, such as nitrate and phosphate. **Eutrophic** is the term applied to a body of water, which has been enriched in this way. Eutrophication can be a natural process, but is often due to the works of man.

Hadal. The ocean deep below 11,000 m.

Halophile. A micro-organism which has an obligate requirement for NaCl at greater than physiological levels. May be sub-classified as moderate and extreme. The term should not be applied to micro-organisms which are NaCl tolerant, but which have no requirement. **Haloalkilophiles** are halophilic micro-organisms which require a high pH for growth.

Heterotroph. An organism, exemplified by animals, which requires organic nutrients.

Humus. The organic fraction of soil, consisting of naturally occurring compounds resistant to degradation by micro-organisms.

Lichen. A composite organism consisting of a fungus living in association with one, or sometimes two, symbionts. Symbionts may either be algae or cyanobacteria.

Lithobiotic. Living in association with rocks and stones. **Endoliths** live in the interior of stones and **epiliths** on the surface.

Methanogens. One of the constituent groups of the archaebacteria. Anaerobic micro-organisms which convert fermentation products of other anaerobes to methane.

Methophils. A group of eubacteria having the

ability to use single carbon compounds, including methane, as its sole source of carbon and energy. Previously known as methylotrophs.

Mineralization. The decomposition of plant and animal remains and excretion products to inorganic material, which may be utilized by photosynthetic organisms.

Mycorrhiza. Composite structure formed by higher plant roots in symbiotic relationship with a fungus.

Myxobacteria. The distinct group of eubacteria having a gliding motility and, in many cases, a special developmental cycle which involves formation of a fruiting body.

Neuston. The layer at the surface of bodies of water.

Nitrate respiration. The use of nitrate in place of oxygen as the terminal electron acceptor in the growth of some aerobic eubacteria.

Nitrification. The geochemical process in which specialized eubacteria oxidize ammonia to nitrite and nitrite to nitrate. An important part of the nitrogen cycle.

Oligotroph. A micro-organism capable of growth at low nutrient levels.

Parasitism. A symbiotic relationship in which benefit is obtained by one partner only, the parasite. The second partner often suffers damage.

Pelagic. Relating to the open sea. The term is sub-classified as **epipelagic** (surface waters above thermocline), **mesopelagic** (waters directly below thermocline), **bathypelagic** (waters at depth *ca.* 2,000 m) and **abyssopelagic** (water at depth below 3,800 m).

Phototroph. A photosynthetic organism, either prokaryotic or eukaryotic.

Plankton. Floating micro-organisms in aquatic habitats. The term is not restricted to unicellular algae, planktonic bacteria may be described as **bacterioplankton**, ultramicrobacteria as **picoplankton** and bacteriophage as **femtoplankton**. **Zooplankton** describes free floating micro-organisms classified with the animal kingdom. **Phytoplankton** describes all photosynthetic planktonic organisms.

Prokaryote. Organisms in which the unit of structure is the relatively simple prokaryotic cell. These are members of two microbial groups, the eubacteria and the archaebacteria.

Prosthecate. Bacteria characterized by the formation of cell extensions (prosthecae). These are filiform or conical, the interior being continuous with the cytoplasm of the main part of the cell.

Rhizosphere. Regions of soil immediately surrounding plant roots together with the root surfaces of that plant.

Saprophyte. An organism which obtains nutrients directly from dead, or decaying, organic matter.

Symbiosis A strategy for meeting biological competition by adapting to life in close and continuing association with another form of life. Partners are known as symbionts, or bionts, e.g. a mycobiont being a mould partner in a symbiosis and a photobiont, a photosynthetic partner.

Stratification. Formation, in bodies of water, of horizontal layers of distinct chemical and/or physical properties. The boundary is known as the thermocline or chemocline, and stratification may either be temporary (**holomitic**) or permanent (**meromitic**).

Sulphate-reducing bacteria. Strictly anaerobic eubacteria, capable of anaerobic respiration utilizing oxidized S compounds as the electron acceptor.

Acknowledgements

The authors gratefully acknowledge the help of many persons during the writing and preparation of this book. Particular thanks are due to:

Debbie and Phil Andrews for producing many of the hand-drawn and computer-generated drawings and diagrams.

David Post of Oxoid Ltd, for the gift of selective media.

Nikon (UK) Ltd for the loan of a photographic microscope.

The libraries of the BBSRC Institute of Food Research, Reading Laboratory and the University of Reading for their help in obtaining information.

Our colleagues, both in the UK and overseas, who have permitted us to reproduce illustrations derived from their research. *Individual acknowledgments accompany the relevant caption.*

Colleagues, both in Reading and elsewhere, for their help and interest during the preparation of the book.

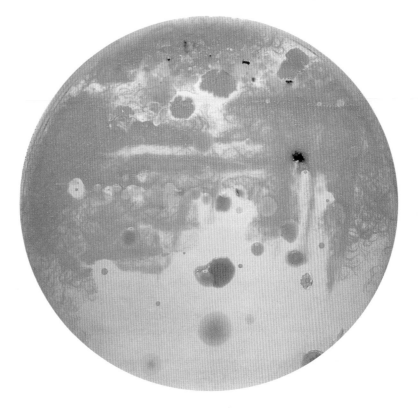

1 OVERVIEW

Until the 1960s many aspects of environmental microbiology were dealt with on an observational basis. Certainly there was a good measure of understanding of economically important phenomena, such as the symbiotic relationship between *Rhizobium* and leguminous plants, while there was a basic understanding of the relationship between microbial growth and physical parameters, such as temperature and of the various types of interaction between different micro-organisms in the same environment. Overall, however, many observations remained poorly explained and, to many microbiologists, it appeared that the various paradigms used to explain and interpret the behaviour of micro-organisms in natural environments were deficient. Over the past two decades, the situation has changed and while, happily for environmental microbiologists, much remains less than fully understood, there now exists a framework of knowledge which has greatly expanded the understanding of the relationship between micro-organisms and their environments. Two examples are an understanding of biofilm formation and its environmental importance and of the survival strategies of non-endospore forming micro-organisms.

THE NATURE OF MICROBIAL COMMUNITIES

Introduction

Natural environments are usually inhabited by a diverse population of micro-organisms. These can encompass a wide range of physiological and nutritional types, from autotrophs to heterotrophs, from psychrophiles to hyperthermophiles and from obligate aerobes through micro-aerophilic bacteria to oxygen-sensitive anaerobes. There are exceptions to this general rule, usually where the environment has been manipulated by man or where the extreme physicochemical nature limits the types able to grow. Examples of the latter include highly alkaline and saline lakes and highly acidic hot springs. Even under these extreme conditions, however, diversity can be much greater than was thought possible in earlier years – in recent years many other concepts in environmental microbiology have been replaced. It was previously considered, for example, that micro-organisms dwelt in discrete compartments for each physiological and nutritional type but it is now recognized that most microbial ecosystems are heterogeneous. More recently still, advances in methodology, in the framework of challenges to earlier orthodoxies, have gone a considerable way to elucidate the development and function of such diverse populations.

Spatial organization

It is now accepted that the indigenous microflora of most environments develop in spatially organized physicochemical gradients (Keith *et al.*, 1987); the distance over which these gradients are generated varies very considerably. Microbial films (biofilms) may be only a few micrometres thick in low-nutrient environments, while soil crumbs, microbial mats, sewage flocs and biofilms in nutrient-rich environments are of macroscopic rather than microscopic dimensions. Less commonly, gradients may extend over many metres. Such situations, which include stratified lakes, sewage outfalls and geothermal springs, can be seen as exceptional but important environments.

It is the existence of a physicochemical gradient that permits the development and coexistence of a heterogeneous population of micro-organisms. The microbial population is organized either horizontally or vertically depending on the direction of the gradient. In many environments a wide variety of physiological and nutritional types are present and micro-organisms with apparently conflicting lifestyles can grow in close proximity. This is a particularly marked phenomenon in physically small environments where the gradients can be very steep. The oxygen gradient across small soil particles, for example, permits the growth of aerobes, micro-aerophils and anaerobes within a physically very short distance.

Heterogeneous communities are of major importance in the microbial world because of the considerable advantages gained by members of the population. Indeed it has been stated that spatial organization in biofilms and similar situations permits micro-organisms to obtain many of the benefits of multicellular life. Interaction between micro-organisms permits activities such as co-metabolism and cross-feeding, while diverse populations are less affected by environmental change and can recover from disaster more rapidly than ecosystems of lower diversity. This leads to long-term stability, although it should be appreciated that most microbial ecosystems are dynamic communities and are subject to continual short-term changes. Changes can result from either periodic or non-periodic events affecting either the physicochemistry of the environment as a whole, or the gradients within a given environment. The physicochemical effects are both direct (through the immediate effect on a given part of the population) and indirect (through the effect on interactions between members of the community). It may be considered, however, that short-term changes enhance long-term stability.

Biofilms

A number of definitions of biofilm have been proposed. The most satisfactory is considered to be that 'a biofilm is a community of microbes embedded in an organic polymer matrix, adhering to a surface' (*Feature 1*).

Feature 1. Biofilms and man

In all known habitats, bacteria preferentially reproduce on a surface rather than in suspension in the liquid phase. Biofilms were first extensively studied during the 1940s but it was not until the 1970s that it was appreciated that their formation occurs in almost all natural environments. 'A rock immersed in a stream, an implant in the human body, a tooth, a water pipe or conduit, etc. are all sites where biofilms develop' (Carpentier and Cerf, 1993). Given the extent to which biofilms impinge on man and his works, the belated recognition of their full importance is rather surprising. From the medical viewpoint, colonization of implants can serve as focal points for potentially fatal infections, the cells in deep layers being protected from antibodies and phagocytic white blood cells (Anwar and Costerton, 1992). Biofilm formation is also a key factor in the pathology of dental disease. In the food industry, it is thought that development of biofilms on machinery surfaces serves to protect pathogens such as *Listeria monocytogenes* from sanitizing agents, while in other industries biofilms are responsible for substantial economic loss (*Table 1*). The world of microbiology is never simple of course and biofilm formation can be exploited by man. *Acetobacter* spp., growing in a biofilm for example, are responsible for the oxidation of ethanol to acetic acid in traditional vinegar manufacture. More commonly, in modern industrial applications, exploitation of biofilms involves their high efficiency in entrapping molecules and small particulate matter and the development of consortia able to degrade even highly recalcitrant molecules (*Table 2*).

Table 1. Examples of industrial problems caused by biofilm formation.

Industry	Problem
Fluid transfer (general)	Reduction of flow rates and blocking of pipes
Potable water distribution	Reduction of flow rates
	Taints
	Unacceptably high bacterial numbers
Paper making	Poor quality paper
Flour milling; malting; sugar refining	Interference with operation through slime formation
Ships' hulls	Increase in fuel consumption; lower overall speeds
Structural steelwork, pipelines, etc.	Accelerated corrosion
Heating and cooling operations	Reduced efficiency
Food processing	Reservoir of spoilage and potentially pathogenic micro-organisms; possible survival of pathogens through under-processing
Printing	Blockage of ink supply pipes; reduced ink quality

Table 2. Processes involving the exploitation of biofilm formation, with some examples.

Process	Example of exploitation
Water purification	Slow sand filtration
Sewage purification	Percolating filters
Agricultural waste purification	Floccor towers
Industrial waste purification	Bioreactors
Removal of heavy metal pollution (including uranium)	Mercury waste remediation
Recovery of metals from low-grade ores	Microbial leaching
Production of acetic acid from ethanol	Traditional process

A number of physiological changes occur in micro-organisms immediately after adhesion, these changes increasing in magnitude with time. Most biofilms in natural habitats consist of an assemblage of different micro-organisms and are heterogeneous in both structure and chemical composition. In all cases, however, the major constituent of the matrix is water. The remainder consists of various extracellular polymers including glycoproteins and polysaccharides, the nature of which is poorly understood. These are usually referred to as extracellular polymeric substances (EPS) and comprise 50–95% of the dry weight of the biofilm. In the past, it was thought that micro-organisms were more or less evenly distributed throughout the biofilm. This may be the case in some circumstances but the usual structure is a collection of micro-colonies, separated by 'water channels' (**1**, **2**).

Biofilms confer on their inhabitants the benefits of a stable environment enjoyed by members of most heterogeneous microbial communities. Biofilms also serve to physically protect microbial cells from external factors such as ultraviolet (UV) irradiation and heat, and to minimize the effects of changes in physical parameters such as pH as well as offering protection from inhibitory substances. Protection from inhibitors can be particularly important where microbial growth is controlled by biocides and disinfectants but it is also of considerable significance in permitting the survival of bacteria following severe pollution.

In addition, the entrapment of water by biofilms minimizes the risk of dehydration. This is of particular significance to micro-organisms in aquatic

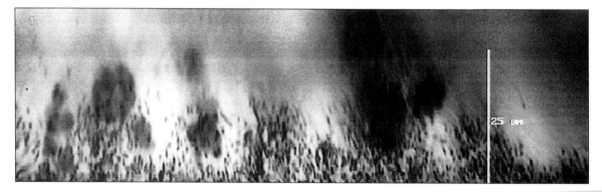

1 Microcolonies within a biofilm developing on the surfaces of a glass flow cell. Images obtained in cross-section show microcolonies forming individual mounds of cells.

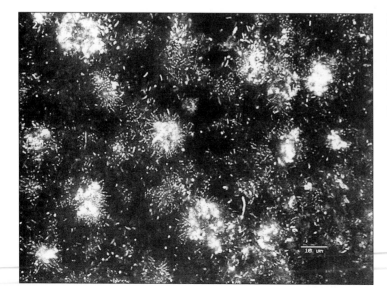

2 Microcolonies within a biofilm developing on the surfaces of a glass flow cell. Images obtained through the film show microcolonies forming individual mounds of cells. **1** and **2** reproduced with permission from Woolfardt G.M., *et al.*, 1994, © 1994 The American Society for Microbiology.

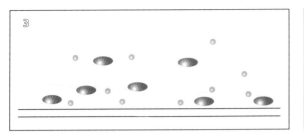

3 Diagrammatic representation of biofilm formation. Initially both bacteria (red) and particulate material (green) are suspended freely above the surface to be colonized.

4 Following **3** there is a concentration to the surface.

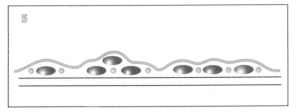

5 Following **4** there is the formation of the 'young' biofilm.

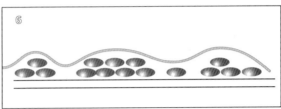

6 As cell multiplication continues, the biofilm thickens to the point at which 'sloughing off' occurs.

environments, periodically exposed to drying-out, as well as in soils, which are subject to seasonal dehydration. As well as protection from physical and chemical stress, biofilms shield bacteria from phages (Hicks and Rowbury, 1987). It may also be that there is protection from lytic and parasitic bacteria as well as from eukaryotic predators. At the same time, predatory micro-organisms may well benefit from the high population of bacteria in the vicinity of biofilms as a result of sloughing off of the outer layers and the release of daughter cells.

Nutrient diffusion through biofilms tends to be slow and only limited quantities are available to micro-organisms in the deep layers. Metabolic activity by such cells is limited and their status has been described as 'quiescent' (Lewis and Gattie, 1990). This status does have the advantage that the bacteria are not subject to competition. At the same time, biofilms are highly effective in trapping nutrients in the form of organic and mineral molecules or discrete particles. This results from the overall negative charge on the polysaccharide molecules within the biofilm matrix.

The formation of biofilms follows a defined sequence, differing on chemically inert surfaces and in living tissue. The fundamental difference lies in the fact that only non-specific adhesion is involved in the environment, while specific adhesion, by means of lectins and adhesins, is a basic process in the development of biofilms in living tissue though only the development of biofilms in the natural environment will be discussed in detail here.

Biofilm formation on chemically inert surfaces involves four stages: transport, adsorption of molecules, adhesion of micro-organisms and colonization (3–6). Transport involves the movement of molecules, inert particles and micro-organisms to the surface on which the biofilm subsequently forms and involves several processes. This is followed by adsorption of molecules, which occurs almost instantaneously. The accumulation of molecules at the solid–liquid interface may be referred to as the 'conditioning film'. Microbial growth rates are enhanced by the higher nutrient concentration at the interface. Adhesion occurs much more slowly than adsorption and involves two distinct stages. In the first, reversible, stage the bacteria continue to show Brownian movement and are easily removed by rinsing or other simple mechanical procedures. The second stage is regarded as irreversible and involves much stronger binding. The degree of adhesion of individual types of micro-organism is determined by the surface properties which, in turn, are determined by the EPS. There is also variation according to the phase of growth of the cell. Exceptions are known, but

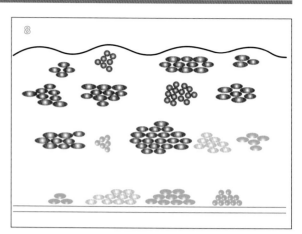

7 Diagrammatic representation of nutrient limitation in biofilms. Even in a relatively newly established biofilm some cells in the lower layer become 'quiescent' (coloured yellow in diagram).

8 As the thickness in **7** increases some colonies in the lower layers become moribund (coloured green in the diagram) while an increasing number move to the 'quiescent' state.

in the majority of cases, EPS released at the end of the exponential growth phase, or during the stationary (starvation) phase, favours adhesion to hydrophobic surfaces. This is consistent with the finding that, in a marine environment, submerged surfaces are usually first colonized by small cells, a small cell size usually being associated with starvation (Roszak and Colwell, 1987). There are obvious advantages for starving bacteria in being primary colonizers in the area of relatively high nutrient concentration at the solid–liquid interface. Extrinsic factors, such as growth temperature may, however, also be involved. Colonization is the final stage and leads directly to formation of the biofilm. Organisms producing EPS multiply within the developing matrix, ultimately forming microcolonies. The rate of colonization varies considerably and the time taken to reach an equilibrium state ranges from a few hours to several months.

A biofilm is a dynamic rather than a static entity, which inevitably undergoes changes with time. The outer surface is continually changing as cells both peel away and adhere. Cells which peel away are usually the more hydrophilic daughter cells. Biofilms can, therefore, be seen to provide a continuing inoculum of daughter cells. This is of environmental importance in that cells, adapted to a given environment, are available to initiate biofilm formation on new surfaces, such as a branch fallen into a stream, or on existing surfaces which, for some reason, have not previously been colonized.

Despite sloughing off, biofilms tend to increase in thickness, leading to changes in the physicochemical gradient. Events within the biofilm can be affected by many factors, including changes in nutrient and O_2 availability, and pH. In the case of nutrients, cells close to the surface of the biofilm receive sufficient to support reproduction, in contrast to the situation in deeper layers where nutrient levels may only be sufficient for maintenance, or even lower. As the thickness of the biofilm increases, conditions within the sub-surface layers change (**7, 8**). The most obvious effect is that the availability of nutrients is decreased leading to previously active cells becoming quiescent. This effect is not absolute and cells of bacteria of lower nutrient requirements, or which are more readily adaptable to alternative substrates, will continue to reproduce for longer periods, thus changing the balance between different species. At some point, however, the thickness of the film becomes such that the nutrient supply in the deepest layers is insufficient to supply even maintenance requirements of the quiescent bacteria. Cell death ensues, leading to the film becoming fragile and sloughing off from the substrate. Biofilm thickness increases most rapidly during periods of high nutrient availability which are favourable to the actively growing cells in the outer layers of the biofilm. Film stability may, however, be reduced by the increasing thickness leading rapidly to the point at which nutrient diffusion to the deeper layers effectively ceases and the death of the microorganisms in those layers.

The effect of stress and disaster on microbial communities

All microbial communities are subject to stress as a result of changes in environmental conditions significantly outside normal parameters. On some occasions stress may be sufficiently severe to cause a disastrous (destructive) effect on the community involving the death of all, or a high proportion, of its population. Stress may result from natural changes or from the works of man. Many of the more dramatic changes are attributed to man and may be deliberate (e.g. sterilization of food) or accidental (e.g. release of pollutants with strong anti-microbial activity). Dramatic changes can also be caused by natural events such as wildfires.

Stress has been defined in various ways, but a practical definition is 'an abiotic factor or set of factors which limit the production of biomass'.

Stress factors are typically unfavourable temperature, pH value, nutrient supply, water availability, and the like. Stress does not, of course, affect all members of a population equally and as a general rule normal parameters are much narrower for eukaryotic organisms than prokaryotic.

The effect of stress obviously depends on the nature, extent and rapidity with which change occurred. Caution is required, however, in that small changes in environmental conditions can have much greater impact than predicted. Where change is slow, members of microbial populations can, within limits, adapt (habituate) to changing environmental parameters and changes over protracted periods may have relatively limited effects (*Feature 2*).

Feature 2. Storm and stress

Concerns over undesirable and possibly dangerous environmental changes are, with justification, a preoccupation of the 1990s. We tend, however, to think in terms of direct threats to man and the macro-environment. Examples are the connection between the diminution of the ozone layer and skin cancer, and deforestation due to acid rain. It is less commonly recognized, even by environmentalists, that these and other factors also place considerable stress on microbial communities. Such stress may affect man indirectly, through loss of productivity and consequent effects on the food chain to higher life, or directly through loss of desirable properties exploited in activities such as bioremediation. Micro-organisms do, however, have a remarkable capacity to adapt and this may be illustrated by a brief examination of a potentially disastrous situation: acidification.

Acidification of water and soils in some parts of the world has led to serious problems. Amongst the best publicized is the damage to forest trees caused by acid rain or snow, at least partly a consequence of the burning of high-sulphur coal in power stations. Acidification may also result from agricultural practices and industrial pollution. The slow development of acidification means that a high level of induced tolerance (habituation) is possible. This permits micro-organisms to survive in acidic environments which would rapidly be lethal to their non-habituated counterparts. Acid tolerance may also induce cross-tolerance to other types of stress. It is thought that phosphate may have a role in enhancing ability for survival and growth at sub-optimal pH values. This is of obvious consequence in high-phosphate environments.

Adaptation has often been considered a property restricted to Gram-negative bacteria but it has also been demonstrated for a few Gram-positive genera, including *Listeria monocytogenes*. Even amongst the Gram-negative genera there is considerable variation. A very high degree of adaptation is possible for *Citrobacter*, for example, but very little for *Serratia*.

Although acidification is generally considered to arise from human activity, natural causes, such as the emission of sulphur dioxide during volcanic eruptions, can also be involved. Archaeological studies in the north of Scotland in 1996 produced evidence of a stable neolithic community destroyed as a result of crop failure. This is thought to have been due to acid rain following a major volcanic eruption. Data concerning adaptation is based on Nojoumi *et al.*, 1995.

Despite adaptation, stress is often marked by a decrease in diversity. This may be seen in the eukaryotic population, but not in the prokaryotic, but in extreme cases the population is reduced to a single dominant species. This is seen within foods where processing is manipulated to reduce the number of potential spoilage organisms. Equally, however, highly stressful environments may be exploited by specialized micro-organisms (such as thermoacidophiles) as a means of minimizing competition. In such situations microbial productivity can be very high in the absence of mutual competition between members of a diverse community and consequent reduction of growth rates. This phenomenon may also be seen when apparently favourable changes result indirectly in stress through increased competition. Increased availability of nutrients, for example, which would enhance growth of most members of a community in pure culture, can lead to a reduction in diversity during aquatic blooms as a consequence of competition stress.

In natural environments recovery from disasters often involves re-colonization from unaffected areas. Immediately after the disaster, however, surviving organisms, often endospore-forming bacteria, are able to dominate the microflora. The microflora usually remains in a non-equilibrium state during initial stages of re-colonization. In many cases, cyanobacteria are the primary colonizers of denuded areas. Ultimately, however, a stable community will be re-established. The relationship of this community to that existing before largely depends on the extent to which the physicochemical nature of the environment has changed. Even where the physicochemical conditions return rapidly to those pertaining before the disaster, however, re-establishment of the microflora can be a very slow process. Under some conditions, the reasons for this are obscure.

Definition of the structure and diversity of the constituent population

Defining the structure and diversity of natural microbial communities through quantification of the constituent population has long been seen as one of the major objectives in environmental microbiology. Conventional approaches invariably involve culture followed by some form of classification, including the identification of isolates according to prevailing taxonomic criteria, determination of physiological or biochemical properties, or simply according to morphology, etc. Such approaches suffer from severe limitations in that conventional culture techniques have been estimated to recover only between 1 and 10% of those micro-organisms present, although some microbiologists believe that such poor recoveries are a consequence of inadequate technique. It has, for example, been possible to culture the dominant bacteria in seasonal blooms in near-shore waters (Rehnstam *et al.*, 1993). Many problems appear to stem from the continuing use of rich culture media which select for copiotrophic micro-organisms rather than the oligotrophs present in many natural environments. Despite these difficulties, cultural methods are still of use in many situations providing that the inherent limitations are fully understood.

Definitive techniques based on non-cultural methods are, however, enabling the understanding of the environment to be taken beyond the boundaries imposed by the limitations of traditional techniques. Microscopy has undergone something of a renaissance, with the application of new techniques enabling sophisticated visualization of the morphology of micro-organisms in their environments to be made. Assay for characteristic cellular constituents has also been a widely used approach. Fatty acid analysis is one of the most successful of these techniques, allowing for the estimation of biomass and, by use of biomarkers, determination of the incidence of specific groups of micro-organisms (Rajendram *et al.*, 1994). The greatest overall impact, however, has been made by genetic techniques.

Genetic techniques are, of course, well established. In the early days, when the use of guanosine plus cytosine (G+C) ratios was providing a platform to underpin phenotypic taxonomy, genetic techniques appeared to be solely in the realm of the taxonomists. Later, genetic techniques were applied to detection of specific micro-organisms, usually pathogens, in medical, and subsequently, food microbiology. DNA hybridization was widely used, being designed either to detect a whole organism or probes to detect genes encoding a specific factor – usually, in medical bacteriology, virulence. An important development was the polymerase chain reaction, in its various forms, which offers a highly sensitive means of detecting both live and dead cells, a feature which can both be an advantage and a disadvantage.

Genetic techniques are widely used in environmental microbiology, the number and diversity of applications possibly being the greatest of any branch of the science. Both probes and the polymerase chain reaction have been used in studies of microbial communities, although currently the most powerful tool in many applications is the determination of 16S rRNA or 16S rRNA F1 encoding DNA (rDNA). A 16S rRNA probe was used to determine evidence for indigenous *Streptomyces* populations in a marine environment, for example, (Moran *et al.*, 1995), but the capabilities of the tool were perhaps demonstrated more fully by Moyer *et al.* (1994). These workers were able to estimate the diversity and community structure of a microbial mat at an active hydrothermal vent system by determination of variations in 16S rRNA genes. This would simply not have been possible without genetic techniques.

It is inevitable that genetic techniques will be further developed. Development is a two-way process: as our knowledge of the environment expands, we require increasingly sophisticated tools to explore further and as those sophisticated tools are developed we see new opportunities for their application. There is only one note of caution: however valuable any technique is, it is not a philosopher's stone. Every technique has its limitations and it is necessary to be aware of, and understand, these if application is to be successful.

LIFE AT LOW NUTRIENT CONCENTRATIONS: THE COMMON LOT OF MICRO-ORGANISMS

The general use of nutritionally rich media for *in vitro* cultivation of micro-organisms can lead to a misunderstanding of the nutrient status of natural environments. In most natural environments, oligotrophic conditions are the rule rather than the exception and a continuing state of starvation is the lot of many micro-organisms. Even in environments such as the animal body, which appear as a 'horn of plenty' for micro-organisms, nutrient deficiencies may limit growth. Iron deprivation, for example, is an important factor in limiting growth of invasive micro-organisms and reduction in plasma iron levels (hypoferraemia of infection) is an equally important defensive response to bacterial invasion. Some foods do represent a nutrient-rich environment but, as in many artificial media, metabolism is profligate and growth of individual bacteria becomes limited both by deficiencies in key nutrients and by build-up of metabolites.

In many cases the majority of bacteria present are heterotrophic (chemo-organotrophic). Heterotrophs in natural environments are usually limited by the availability of carbon; the rate of carbon-associated energy flow ultimately determining multiplication, maintenance or death of cells. At low carbon levels, energy requirements are growth rate dependent, in contrast to the growth rate independent model for high carbon levels (Muller and Babel, 1996). Available energy is always partitioned between growth and maintenance. Maintenance processes have a higher affinity for ATP but cannot command the total energy supply (Chesbro *et al.*, 1979). It should be appreciated, however, that while carbon is the common limiting factor for heterotrophs, low levels of inorganic phosphates may also limit population development (although under some circumstances phosphates are present in abundance). Phosphate limitation is an example of starvation stress and is a particular phenomenon of natural waters, although inorganic phosphate limitation can occur elsewhere. Paradoxically, these environments can contain high levels of organic phosphates derived from biological or anthropomorphic sources, but utilization requires the micro-organism to be able to synthesize phosphatases.

This ability appears to impart a significant competitive advantage (Ozkanka and Flint, 1996).

In environments such as water and soil, heterotrophic micro-organisms may be classified as oligotrophs and copiotrophs. Oligotrophs are characterized by a low growth rate, highly efficient substrate scavenging and accumulation of nutrients, over a long period, to permit multiplication. This lifestyle is the norm even at relatively high nutrient levels. Copiotrophs, at high nutrient levels, are of low substrate affinity and high growth rate. At limiting nutrient levels, however, copiotrophs use a number of survival strategies. Complex molecular regulatory mechanisms are activated in response to starvation. Expression of starvation genes results in a reduction in size (**9**) and the formation of 'ultramicrobacteria', usually by a process of reductive division (**10, 11**). Protective 'starvation proteins' are synthesized and substrate capturing enhanced. Amino acid uptake, for example, switches from low affinity to high affinity in response to nutrient limitation. In at least some cases, such cells are culturable providing sufficient care is taken (**12**). All bacteria, however, have a point at which exogenous nutrients can no longer be obtained and are thus effectively exhausted. At this point cells may enter a dormant, anabiotic state that permits them to survive for long periods without division (Kaprelyants *et al.*, 1993). During this state cells are considered to be nonculturable. It also seems likely that in some situations at least true dormancy is preceded by a 'nutritionally opportunistic' state in which endogenous metabolism is sufficient to maintain the ability of the cell to respond very rapidly to nutrient availability. A low level of endogenous metabolism, together with a low maintenance energy requirement, would therefore appear to be of considerable importance not only in determining the capacity of micro-organisms to survive, but also in responding rapidly to the advent of more favourable conditions. *Aeromonas hydrophila* is an example of a micro-organism that can adapt to a wide range of nutrient concentrations. It appears, however, that at low nutrient levels the advantages of this inherent property are often undone by the presence of autochthonous micro-organisms. This is probably a result of competition for the available nutrients (Kersters *et al.*, 1996).

It should be mentioned that observations have solely involved aerobic bacteria and that the situation with obligate anaerobes may be entirely different.

There has been some interest in using ultra-microbacteria in biotechnology. Major interest has involved the use of genetic modification to provide non-multiplying cells which can express a high level of a desired biochemical activity in dense fermenter cultures. A further possibility is use in bioremediation where conventional methods are limited by low nutrient levels leading to low metabolic activity.

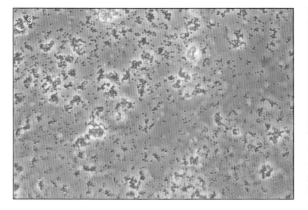

9 Bacterial cells under conditions of nutrient starvation. A marine *Vibrio* underwent a rod-coccus transformation while held in artificial seawater at room temperature for 20 days. About 50% of cells could be recovered by direct plating on to non-selective artificial seawater medium. (Microscopy: phase contrast, × 1,000.)

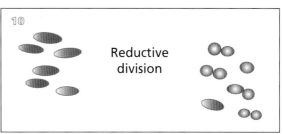

10 At least some bacteria have a viable nonculturable response resulting from a temperature downshift which is distinct from, but related to, the starvation response. This is a strain of *Vibrio vulnificus* that underwent reductive division in seawater at 25°C (77°F) but remained recoverable by standard procedures for a prolonged period. From Oliver *et al.*, 1995.

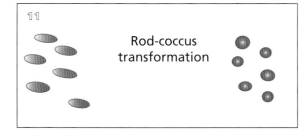

11 In seawater at 5°C (41°F) cells underwent rod-coccus transformation but reductive division was not involved and the cells rapidly become nonculturable. From Oliver *et al.*, 1995.

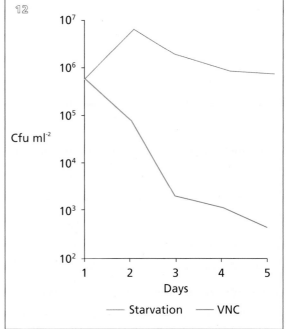

12 Entry into the nonculturable state by *V. vulnificus* in seawater at 25 °C (77°F) and 5 °C (41°F). After Oliver *et al.*, 1995.

PHYSICOCHEMICAL FACTORS AFFECTING THE ENVIRONMENTAL FATE OF MICRO-ORGANISMS

As discussed above, the constituents of microbial communities are determined by the physicochemical conditions and by nutrient availability. These may be seen as determining the broad composition of communities in that similar physiological and nutritional types will be selected at each niche. In many cases dominance within each niche will be determined by the competitive ability of each micro-organism against others selected by the same environmental pressures (see pages 23, 24).

It is common to describe the influence of physicochemical factors on micro-organisms in terms of parameters determined in pure culture. This must be treated with extreme caution, especially since the effect of one parameter is usually determined when other parameters are optimal. Equally it is necessary to be aware, when evaluating the effects of parameters such as pH value, that the physicochemical status of the environment as a whole may be very different to that at the level of the micro-organism. Conventional means of measuring parameters are of little value and the use of microelectrodes is required. These have tip diameters in the region of 50–100 µm although diameters of less than 10 µm are common in the determination of dissolved O_2 (Herbert, 1992).

A great many environmental parameters affect the fate of micro-organisms. Those of greatest *overall* importance will be discussed here (temperature, composition of atmosphere, pH value, water availability and light radiation). It is recognized that other factors, including both those naturally occurring (such as concentrations of salts) and 'artificial' factors (such as added preservatives in foods and other industrial products) are important in some instances, some of which are discussed later. At the same time, the present discussion is largely restricted to non-extreme environments; extreme environments are discussed in Extreme Environments (page 131). Some overlap between extreme and non-extreme environments is, however, inevitable.

It is also necessary to appreciate that the relative importance of the different factors varies according to the environment and the individual micro-organism. It is possible to envisage each factor as a hurdle of varying height (the hurdle concept) each of which must be 'jumped' by a micro-organism to be able to develop in an environment. This concept is useful but does not fully take account of interactions between physicochemical factors. These are very difficult to predict, but advantage may be taken of recently developed mathematical modelling techniques.

A mathematical model is a set of equations representing a prototype (a representation of the system under study). Two main types of model have evolved (Bazin and Prosser, 1992), the first – the mechanistic model – is used to explain how an ecosystem operates. The second – the predictive model – is used to forecast future events.

Temperature

Temperature is demonstrably of very considerable importance in determining microbial growth rates and metabolic activity. In many environments, however, low temperatures are common and it is unsurprising that these habitats have been colonized by cold-adapted micro-organisms, able to grow relatively rapidly at temperatures of *c*.0°C (32°F). Such micro-organisms may be considered to be of two types, psychrotrophs (which are able to grow at low temperatures but have a maximum temperature above 20°C (68°F)) and psychrophiles (which have an optimum for growth of 15°C (59°F) or lower and are unable to grow at temperatures above 20°C). In contrast to very high temperatures, temperatures near 0°C are not considered 'extreme' and a wide range of micro-organisms – comprising representatives of both Gram-negative and Gram-positive bacteria, cyanobacteria, fungi and eukaryotic algae – are either psychrotrophic or psychrophilic. Many are members of genera that are common in higher temperature environments. In contrast to the general rule, however, a number of psychrophilic bacteria have been isolated that do not have direct counterparts in warmer environments. These include '*Micrococcus cryophilus*' (which has a typical Gram-negative cell envelope and is not a species of *Micrococcus*), *Arthrobacter glacialis*, *Aquaspirillum arcticum* and '*Vibrio psychroerythrus*'.

Psychrotrophic micro-organisms are found in a wide range of habitats (**13**) including those that are permanently cold and those where temperatures fluctuate on either a regular or irregular basis. In contrast, psychrophilic micro-organisms are found only in environments that are permanently below 5°C (41°F). The most important of these is marine waters, 90% of the volume of the ocean being below 5°C. Water in the depths of deep mountain lakes is also permanently cold. Permanently cold soils and sediments are found in the polar regions, in caves and in the vicinity of ice-fields. Even in such permanently cold habitats psychrotrophs outnumber psychrophiles. It has been suggested, however, that in at least some cold habitats psychrophiles represent the autochthonous microflora and psychrotrophs the allochthonous flora (Baross and Morita, 1978).

Life at low temperatures requires significant adaptation to maintain replicative ability. The efficiencies of electron transport, ion pumping and nutrient uptake have been described as determinants of psychrophilly and depend on membrane lipid fluidity and protein activity.

Micro-organisms produce stress proteins in response to a shift to low temperatures. The psychrotrophic yeast *Trichosporon pullulans*, for example, produced 26 stress proteins in response to a 21 to 5°C (70 to 41°F) cold shock but only six proteins in response to a 15 to 5°C (59 to 41°F) shock. In the case of micro-organisms in habitats that are permanently cold, proteins necessary for life at low temperatures must be permanently synthesized as a result of cold adaptation. Such proteins must be distinguished from stress proteins which confer a transient protection (Gounot, 1991). Environmental micro-organisms unable to grow at seasonal low temperatures also have survival mechanisms which appear related to, but distinct from, mechanisms for survival at very low nutrient concentrations. A limited degree of adaptation is possible to low temperature growth although the significance is not clear. This is illustrated by the fact that in cold environments the optimal temperature for growth and carbon metabolism is usually very significantly higher than the *in situ* temperature. At higher temperatures there is a much closer adaptation of growth and metabolism to ambient temperature.

Until recently there has been relatively little interest in the use of psychrotrophic and psychro-

13 The importance of low temperatures in extending the storage life of food was recognized empirically many years before systematic studies of the relationship between temperature and microbial growth. Large-scale production of ice at central plants was common before cheap mechanical refrigeration became available to individual shops and the like.

philic micro-organisms in biotechnology. It is recognized, however, that the ability of such micro-organisms and their enzymes to be metabolically active at low temperatures has potential advantages in a number of areas. It has also been recognized that cold environments, especially those which are stressful in other ways, such as low nutrient availability, high salinity and high pH value, may be an important source of micro-organisms with novel enzymes of potential industrial application.

In pure culture, many micro-organisms are less tolerant of temperatures above their optimum than those below. In many natural environments, the effect of temperature increase is very significant and can lead to stress of the population even when increases are small. In many cases this is probably an indirect effect, resulting from increased competition and other factors. Protection against high temperatures can be conferred by stress proteins and limited adaptation is possible both to growth at higher temperatures and to withstanding temperatures in the lethal range.

Atmosphere

An oxygen-rich atmosphere developed on Earth about two billion years ago as a result of the activity of oxygenic, photosynthetic prokaryotes. This event allowed the development of pathways, utilizing oxygen as terminal electron acceptor, which yield high quantities of energy. In aerobic eukaryotes the final stages of oxidative phosphorylation are carried out in mitochondria which are themselves descendants of endosymbiotic, alpha-proteobacteria. Despite this many natural environments are anoxic, or have significantly reduced oxygen tensions. Inevitably these environments support a significant population of micro-organisms, primarily but not entirely prokaryotic, which are not only able to thrive but which may be actively damaged by oxygen.

It is often thought that obligate anaerobes predated the development of an oxygen-rich atmosphere. This is not necessarily true, however, since an anaerobic life style can be the result of secondary adaptation. This may be illustrated by reference to the two groups of eukaryotic, anaerobic micro-organisms. The first of these, the Archezoa, composed principally of trichomonads, diplomonads and microsporidia are 'primitively' anaerobic and considered to be 'living relics' of the earliest phase of eukaryotic evolution when free O_2 was scarce (Cavalier-Smith, 1987). The second group of anaerobic eukaryotes branched from the eukaryotic tree later than the mitochondria-containing Euglenoza, which includes *Euglena* and *Trypanosoma*. Anaerobic eukaryotes of this type are probably secondarily adapted to anaerobic environments. Many, such as anaerobic chytrid fungi and protozoa of the rumen are actually endosymbionts, but a few free-living species are known.

The distinction between anaerobic and aerobic environments is by no means clear cut and a third classification, micro-aerophils, exists to describe those micro-organisms that are aerobic but can only grow under reduced oxygen tensions. The situation is further confused by the widespread existence of facultative anaerobes that can grow anaerobically by carbohydrate fermentation and the means by which non-fermentative micro-organisms can grow in the absence of air. These include nitrate respiration and the arginine dihydrolase pathway. Although environments exist which are clearly the realm of the anaerobes, alternative mechanisms to O_2 respiration enable both obligate aerobes and facultative anaerobes to be highly successful in what might appear to be unfavourable atmospheric conditions. Obligate aerobes such as *Pseudomonas*, for example, have highly efficient, oxygen-scavenging systems and are often able to out-compete micro-aerophilic bacteria in conditions of reduced O_2 tension.

pH value

pH is the key factor in deciding the general outcome of competition between bacteria, yeast and mould in environments which can vary in pH, such as soils and some natural waters. Yeasts and moulds tend to predominate at lower pH values and bacteria at neutral and higher pH values. pH can also be important in the emergence of a dominant strain from a number of closely related micro-organisms. In the natural souring of milk, for example, the number of strains of metabolically active *Lactococcus* falls with falling pH value. Stress selection at low and high pH values means that diversity is limited and under extreme conditions only a small number of genera are able to multiply. Adaptation is possible to both high and low pH values and protection is offered by stress proteins. Very steep pH gradients can exist and extreme care must be taken in applying measurements made using conventional pH electrodes to environments.

Relatively dramatic changes in pH can occur as the result of metabolic activity. Deamination of amino acids and consequent ammonia production leads to an increase in pH value, while fermentation of sugars and acid production leads to a fall. Fermentation acids themselves have an anti-microbial effect which has previously been explained in terms of depression of pH below the growth range and inhibition by undissociated molecules.

This model cannot, however explain differences in sensitivity to acids. It appears that resistant bacteria, in contrast to earlier theories, allow their intracellular pH to decline as a function of extracellular pH. This maintains a more or less constant pH gradient across the cell membrane and prevents the accumulation of high and potentially toxic concentrations of acid anions at low pH (Russell, 1992). Such bacteria have a significant advantage in low pH environments containing large quantities of fermentation acids.

Water availability

The availability of water (often expressed as water activity, a_w) is of major importance to micro-organisms as to other forms of life. Fluctuations in the osmolarity of the environment immediately surrounding an organism are the source of considerable stress and there must be a strategy to permit growth and survival (Booth *et al.*, 1988). Growth requires the maintenance of positive turgor pressure; the pressure differential between the internal osmotic pressure and that of the environment (see definitions of osmolarity and osmotic pressure).

Osmolytes are involved in maintaining turgor. In Gram-negative bacteria, potassium ions are the major cytoplasmic osmolyte, in the absence of another osmoprotectant. The intracellular potassium concentration remains relatively constant, provided that the osmolarity of the environment remains unchanged. As the osmolarity of the environment increases, potassium is taken up by the cell to maintain turgor by raising the intracellular osmotic pressure. In the reverse process, there is an efflux of potassium when turgor becomes high due to a reduction in the osmolarity of the environment. Two sets of genes, the *Kdp* and the *Trk* systems are responsible for potassium uptake and at least three sets of genes for potassium efflux. Less, however, is known of the mechanisms of potassium efflux than those of potassium uptake.

In addition to the role played by potassium, it is now recognized that certain organic solutes, 'compatible solutes', when present in the cell in high concentrations, can significantly enhance growth rates in an environment of high osmolarity. Glycine betaine (*N,N,N*-trimethylglycine) is a compatible substrate commonly found in eubacteria and plays a key role in the osmoregulation of enteric bacteria. Accumulation may result either by uptake from the environment immediately surrounding the cell, or by *de novo* synthesis from the precursor choline (Fernandez-Linares *et al.*, 1996). *De novo* synthesis does, however, appear to be an unusual trait among chemoautotrophic eubacteria. The ability to take up external glycine betaine can be explained, in the case of *Escherichia coli* at least, by its inability to oxidize choline anaerobically and thus evolving the ability to scavenge for the solute in the environment. Other compatible solutes are of importance, primarily proline and trehalose. Synthesis of trehalose certainly occurs when potassium levels become extremely elevated in response to very high

> **Definitions**
> When a given solution is separated from a solute-free solvent, which is permeable to the solvent but not the solute, the solvent tends to migrate through the membrane into the solution, with consequent dilution. The hydrostatic pressure required to prevent the migration of solvent is the *osmotic pressure*. The osmotic pressure exerted by a given solution can be defined in terms of *osmolarity*. An osmolar solution is one that contains one *osmole* per litre of substrates.

external osmolarity and is accompanied by a reduction in intracellular potassium levels.

Other compatible substrates also effect a reduction in the intracellular potassium levels. While the trinity of glycine betaine, proline and trehalose is of overwhelming importance in the great majority of eubacteria, the situation is different in that ectoine (1,4,4,6-tetrahydro-2-methyl-4-pyrimidine carboxylic acid) is the compatible solute most commonly synthesized by halophilic, aerobic, heterotrophic bacteria, the majority of which belongs to the gamma subclass of the *Proteobacteria* (Fernandez-Linares *et al.*, 1996).

The means by which compatible solutes exert their protective effects is now well documented. Firstly, the solutes are able to stabilize enzyme activity in solutions of high ionic strength by inducing preferential hydration of proteins in the presence of salts. Secondly, the accumulation of compatible solutes in the cell leads directly to the release of potassium ions and a lowering of the cytoplasmic ionic strength (Sutherland *et al*, 1986). A synergistic relationship exists between the direct effect of betaine and proline on stabilizing protein structure and the indirect effect of lowering ionic strength (Booth *et al.*, 1988).

One of the intriguing facts concerning turgor regulation is that cytoplasmic ionic strength provides the signal for the induction and, possibly, the activity of systems that mediate the synthesis and/or accumulation of compatible solutes. Equally the external osmotic pressure, through reducing turgor, is responsible for differential gene expression. Overall, it may be said that an effective response by the cell to osmotic pressure involves highly integrated functioning of all the elements of the viable cell, regulation of transport and metabolic systems being of particular importance.

Although some environments have permanently abundant supplies of water, others are permanently arid and select a specialized microflora. A specialist microflora is also stress selected in foods and other materials preserved by drying; this usually comprises moulds. A common environmental situation is periodic lack of water due to climatic conditions. This requires survival rather than growth and micro-organisms produce a variety of desiccation-resistant bodies including both endospores and exospores, many of which can persist for extended periods of time. Vegetative cells often gain protection from biofilms but can be remarkably desiccation resistant without any apparent protection. Coccal-shaped cells are more resistant than rod-shaped and drying-out of an environment can change the balance of the population after rehydration.

Light radiation

Terrestrial and aquatic micro-organisms are exposed to both visible and UV radiation. In some circumstances, such as survival of enteric micro-organisms in seawater, light of both visible and UV wavelengths is a major factor. It has been hypothesized that the underlying cause is oxidative stress (Goumelon et al., 1994). Susceptibility to photo-inactivation depends both on species and individual strains, many pigments having a significant protective role. Photosensitizers may be endogenous within the cell, while exogenous photosensitizers, such as humic substances, can also be involved. Susceptibility, especially to UV radiation, is usually significantly greater in micro-organisms from human and animal sources, where exposure is minimal, than in similar micro-organisms from the external environment. This may be demonstrated by reference to *Leptospira*, the free-living *L. biflexa* being far more resistant than the parasite *L. interrogans*. Resistance to UV radiation is, however, highly conserved in subsurface micro-organisms, which are effectively shielded.

Light is, of course, essential to primary production through photosynthesis. Higher plants are adapted to use the highest intensity of light available, but the exposure of most algae and cyanobacteria to light of full intensity would lead to rapid bleaching, loss of photosynthetic ability and ultimately death. Sensitivity to UV-B (280–315 nm) is greatest and populations can be under stress even at normal levels. Increased UV-B irradiation, for example resulting from the thinning of the stratospheric ozone layer, has decreased primary production by phytoplankton in Antarctica

by 6–12%. Decreases in primary production have been noted elsewhere. A number of strategies have been developed amongst these micro-organisms to avoid harmful effects while gaining greatest benefit from exposure to light. Some, for example, produce 'sunscreen' compounds which absorb UV-B and thus reduce the likelihood of damage to target cellular compounds. An alternative strategy involves motility and a movement to the zone of optimum light intensity, where conditions permit photo-synthesis without reducing growth. An exception is the red freshwater flagellate *Euglena sanguinea* which contains photoprotective pigments and which orientates itself to light (Gerber and Hader, 1994). This enables the flagellate to take advantage of high light intensities and the stress-induced lack of competition at the water surface.

No strategy for minimizing exposure to UV radiation, however, can be entirely successful in preventing a degree of damage. The damage primarily involves DNA, which absorbs UV light, the energy so released causing dimerization between adjacent pyrimidine residues on the same DNA strand. The occurrence of the pyrimidine dimers triggers a repair mechanism which excises them, exposing a short region of single-stranded DNA on the opposite strand. Single-stranded regions are also formed if a replication fork passes a pyrimidine dimer that has not been excised. The formation of the single-stranded regions induces the *SOS system*, a general DNA repair system. This, in turn, induces at least nine enzymes that fill the gaps opposite the single-stranded. The SOS system is rapid acting, but relatively inaccurate, being described as error-prone repair. This accounts for the mutations common after exposure to UV and, of course, the deliberate use of UV as a mutagen in genetic studies. A second repair system involves excising defective sections of a DNA strand and their replacement by insertion of new nucleotides. This system is relatively error free. The relative importance of the two systems varies in different bacteria. Bacteria also differ in overall susceptibility to UV damage. Radiation-resistant bacteria, such as *Deinococcus*, for example, have highly effective DNA repair mechanisms, although it is also possible that the DNA of radiation-resistant bacteria is inherently more stable than that of the less resistant. Research has shown, however, that negative responses to UV irradiation are not necessarily due to exposure to UV-B alone but that interactions with the effects of UV-A (320–400 nm) and 'photosynthetic active radiation' (400–700 nm) are also of major importance (Nielsen and Eklund, 1995).

COMPETITIVE STRATEGIES OF MICRO-ORGANISMS

Competition is often considered to involve micro-organisms that have the same *overall response* to the environmental pressures present and thus share the same niche. Considerable care is needed, however, in using even relatively simple definitions. It is necessary, for example, to emphasize overall response to environmental pressures. It is a common misconception that the outcome of competition between two, or more, micro-organisms is decided by one defining factor, such as pH. Except in the very simplest of environments, however, the most successful organism is that which can respond most effectively to a multi-factorial matrix of stress factors. At the same time, key factors may be unrelated. The outcome of competition between two micro-organisms in a given environment may be determined by whether the benefit gained by one organism by its response to Factor A (e.g. efficiency of nutrient utilization) is greater than the benefit gained by the other organism in response to Factor B (e.g. low pH value).

In many cases, competition is between micro-organisms which, physiologically at least, are closely related. This is fairly obvious, since two micro-organisms which respond in the same, or similar, ways to environmental stress are likely to compete in the same niche. In some cases, the competing micro-organisms are strains of the same species. In wine-making, using a natural yeast inoculant, for example, the succession of different karyotypes of *Saccharomyces cervisiae* results from the relative competitive ability of each in conditions of falling sugar and rising alcohol content. It is important, however, to avoid anthropomorphism. It may justifiably be argued that the existence of karyotypes able to compete effectively in conditions ranging from the relatively hospitable to the relatively hostile is itself a strategy to ensure that *Saccharomyces cerevisiae* is able to compete over a wide range of environmental conditions. A similar overall strategy has been adopted by a number of micro-organisms. Many moderately halophilic bacteria, for example, can grow over a remarkably wide range of NaCl concentrations, being the dominant organism at some concentrations, but secondary at others (see page 146). Other micro-organisms have two, or more, sets of key enzymes that are induced in response to differences in environmental circumstances, such as reduced temperatures. The ability to adapt to unfavourable conditions, either by production of stress proteins, or by other routes, also permits some micro-organisms to compete, whether successfully or not, outside their normal range.

Competition does not always involve physiologically similar micro-organisms, but care is required to decide which situations are truly competitive. The ability of aerobic and anaerobic bacteria to coexist in soil, for example, was long an enigma for soil microbiologists, since it would be expected that one category would rapidly attain dominance. The advent of mini-probes, which measure physico-chemical parameters over very small areas, demonstrated that very steep gradients can exist between adjacent, micro-scale areas. Thus aerobes and anaerobes are effectively capable of growth side-by-side. This cannot be considered competition unless the situation is complicated by a stress factor affecting both aerobes and anaerobes, such as nutrient deprivation.

True competition between physiologically distinct micro-organisms does exist, an example being that in micro-oxic environments between microaerophilic bacteria and some aerobes. In such situations, it would be expected that microaerophilic bacteria would be dominant. This is not, however, the case when aerobic bacteria, such as *Pseudomonas*, which can compete effectively in a wide range of environments, are present. Strains of *Pseudomonas* inhabiting micoaerophilic environments, such as those of reed beds (see page 78) have remarkably effective cytochrome systems, which are able to scavenge oxygen imparting a competitive advantage.

In an overall appreciation, three main competitive strategies available to micro-organisms have been identified, although many micro-organisms employ a combination of two, or three. In environments such as most soils and natural waters, which are capable of supporting a large population of micro-organisms, the most common is directly competitive, or combative, and involves competition for nutrients, including greater nutritional versatility. Antagonists specific to other micro-organisms, such as

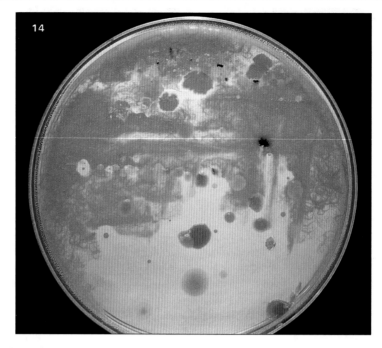

14 Production of inhibitors by some colonies is a common feature of crowded isolation plates from environments such as soil. It *must not*, however, be assumed that such phenomena on artificial media have any significance in the environment. (Source: garden soil; Medium: nutrient agar; Incubation: 22°C (71.6°F), 4 days).

bacteriocins and antibiotics (**14**) may be produced and there may be non-specific antagonism due to environmental modification by, for example, production of fermentation acids. In the case of fungi, there may also be aggressive mycelial interactions.

The second strategy, ruderal, involves the ability of the minority sub-set of the population to respond more rapidly to improved environmental conditions than the current majority sub-set and subsequently attain dominance. The most common environmental condition involved is nutrient status; but other factors, including temperature, pH value and dilution of NaCl levels due to rainfall, can all be of importance. Micro-organisms effective under conditions of ruderal competition are characterized by their rapid uptake of easily assimilable substrates and fast growth rate. Long-term colonization depends on the period over which the changed conditions persist, but is likely to be difficult. That sub-set of the population which is able to take advantage of an improvement in environmental conditions must have a strategy for survival during less favourable times. It has been postulated that the highly heat- and desiccation-resistant endospores of *Bacillus* and *Clostridium* fulfil that role, while fungi and actinomycetes produce exospores which not only are desiccation-resistant, but which can be widely disseminated and thus colonize new areas where conditions are favourable.

It must be appreciated, however, that not all micro-organisms involved in ruderal competition have specific survival mechanisms for unfavourable periods. In some cases, the population falls when nutrient levels, or other favourable factors, can no longer sustain rapid growth and a large population. To be able to take advantage of an improvement in conditions it is probable that a minimum population must be maintained. This may well be accompanied by a marked reduction in metabolic rate and the initiation of the various survival strategies. These, for example, may involve downsizing of cells, usually by reductive division (see page 16), and a switch from low affinity substrate uptake to high affinity.

The third strategy is not really competitive at all in that the organism exploits an extremely unfavourable habitat where competition is effectively non-existent. The best example is that of thermo-acidophiles (see page 133). A considerable degree of physiological adaptation is required, but long-term colonization is possible, albeit in a very restricted number of niches.

INTERACTIONS INVOLVING MICRO-ORGANISMS

Micro-organisms enter a vast range of symbiotic relationships involving one or more partners, which may be other micro-organisms, animals or plants. Micro-organisms can also be a third party in many plant–animal relationships. It should be noted that various definitions are applied to symbiosis – that used here is broadly based and covers any living association no matter how transient (Evans, 1988).

Various definitions have been applied to interactions involving micro-organisms. In some cases it is difficult to determine the nature of the symbiosis. The ectomycorrhizal symbiosis between fungi and plants (see page 108), for example, is considered by some mycologists to be a directed parasitism of either the plant by the fungus or vice versa (Malloch, 1988). More generally the symbiosis has been described as a co-evolved mutualism (Pirozynski and Hawksworth, 1988).

Competition: Competition (see above) is probably the most common interaction, although the number and complexity of mutualistic interactions (see below) may be much greater than previously thought. Where the outcome of competition depends on response to a single limiting factor, the development of both populations is likely to be restricted. In more complex situations, however, where the outcome of competition depends, for example, on the adequacy of response to two different limiting factors, only one population is likely to be restricted. This may lead to total dominance by one population, although co-existence is possible, even when one population is considerably larger.

Amensalism: Amensalism is the condition where the outcome of competition invariably leads to complete dominance by one population, which restricts the growth of its competitor population(s) without itself being affected. Changing conditions can lead to a reversal of the relationship between the populations. This may be drastic, or gradual. Extensive rainfall on highly saline lakes can lead to a dilution that is catastrophic for the previously dominant extremely halophilic archaebacteria and cause dominance by the previously subordinate moderately halophilic eubacteria (see page 147). This reversal is usually fairly short-term. Gradual reversal of the relationship occurs, for example, when a lake is subject to slow acidification (see page 76). The reversal is usually long-term, although there may be further changes in the balance of the microflora if acidification continues.

Commensalism: Commensalism involves benefit to one population while the other is unaffected. A quoted example is the production, by *Arthrobacter globiformis*, of the 'terregens' factor, one of a number of sideramines essential to the growth of soil-dwelling 'coryneform bacteria'. Some microbiologists, however, consider this relationship to be actually mutualistic. This underlines the fact that commensalism is difficult to define and many relationships described as commensal are, in fact, mutualistic or parasitic. Commensalistic relationships are also dynamic and may become parasitic, or mutualistic. Equally, parasitic, or mutualistic relationships may become commensalistic.

Mutualism: Mutualism exists in a very large number of forms. In some, there is mutual dependence and neither micro-organism can grow in the absence of the other. Such dependent relationships are usually stable. Mutualism may, equally, involve no more than a casual relationship, which may be transitory. Mutualism is of considerable importance in the development and maintenance of microbial communities, especially when nutrient levels and other limiting factors for growth are subject to change.

A **microbial consortium** is a stable mixed culture, such as can be found in biofilms. The relationships between members of the consortium are usually, but not invariably, mutualistic.

Parasitism: In both parasitism and the parallel process, predation, one organism benefits directly at the expense of the other. In parasitism the smaller organism usually benefits and infection often leads to the death of the larger host cell. In some cases, it is important to distinguish between the effect of parasitic infection on individual cells and on populations as a whole. Parasitic attack by bacteriophage in the aquatic environment recycles a high proportion of nutrients, which may be utilized directly by surviving micro-organisms.

Predation: In predation, the benefiting organism is usually the larger, although there are exceptions;

the predatory bacterium *Bdellovibrio bacteriovorus*, for example, is considerably smaller than its prey. A number of predatory mechanisms exist and many predators locate prey by means of complex 'signals'. Some aquatic bacteria are known to produce a proportion of large cells amongst the general population and it has been postulated that this is a means of defence against 'small' predators, which do not recognize the prey signal. The large cells are, however, more susceptible to predation by 'large' predators.

Indirect interactions: Interactions are often discussed in terms of micro-organisms impinging directly on each other. This is not, however, correct and many indirect interactions exist, which are of considerable environmental significance. An example is the behaviour of micro-organisms during natural fermentation (involving pH fall) of decaying vegetation. The fermentation is usually initiated by environmental members of the Enterobacteriaceae, acid production, together with a rise in the local CO_2 concentration within the vegetation, creating conditions for growth and further acid production by *Lacobacillus* and other lactic acid bacteria. When the fermentable substrate is exhausted, acid-tolerant, lactate-utilizing yeasts become dominant, the pH value rises and a variety of other bacteria develop following exhaustion of lactate and a fall in the yeast population. Successions of this nature are widespread in the natural environment and in foods. In general, the micro-organism furthest down the succession sequence is the most successful. There are, inevitably, exceptions. *Brevibacterium linens*, which is involved in the ripening of many varieties of surface ripened cheese, initiates changes that permit the growth of lactate-tolerant yeast and subsequently *Penicillium*. Under some, ill-defined, circumstances, *Br. linens* remains the dominant micro-organism despite extensive growth of the yeast and mould. (*Note that this is a fault in varieties such as Camembert and Brie.*)

A further example of an indirect interaction is that between microalgae and aerobic bacteria. Many factors may be involved, but an interesting hypothesis is that the bacteria, (*Pseudomonas* spp.) stimulate the algae (*Scenedesmus* and *Chlorella*) by consumption of photosynthetically produced oxygen. The reduction of O_2 levels in the micro-environment consequently produces ideal conditions for photosynthetic growth. In addition to the benefit obtained by the ready availability of O_2, the bacteria benefit from the release of assimilable carbon compounds by the algae.

Interspecies hydrogen transfer. Methanogenic bacteria have highly efficient hydrogenases. The affinity of these enzymes for their substrate (K_m) is sufficiently high to maintain a vanishingly low partial pressure of H_2 when active methanogenesis is occurring. The resulting low P_{H2} creates conditions which allow a number of fermentative anaerobes to re-oxidize NADH by means of an NADH-coupled hydrogenase:

$$NADH + H^+ \rightleftarrows NAD^+ + H2$$

Although hydrogenases of this type are widely distributed among anaerobes, they are unable to re-oxidize NADH when grown in pure culture, because the equilibrium lies too far to the left. When grown in culture with methanogens present, the constant removal of hydrogen by methanogenesis results in a P_{H2} sufficiently low to shift the equilibrium towards the right and allow the reaction to proceed. In turn, the fermentation balance shifts towards more oxidized products with higher yields of ATP and biomass. Interspecies hydrogen transfer may be seen as a synergistic relationship.

Relationships among a wide range of micro-organisms are discussed in later pages. These include the lichen, mycorrhizal and nitrogen-fixing symbioses, the latter two, in particular being of considerable economic as well as environmental importance (see pages 108–111). Although the examples of economically important symbioses discussed in detail concern plant:micro-organism interactions, the importance of animal: micro-organisms should not be forgotten, although considered beyond the scope of this book. These include the role of micro-organisms in the health and illness of man and other animals and the economic importance of the rumen symbiosis in cattle and sheep.

REFERENCES

Anwar, H. and Costerton, J.W. (1992) Effective use of antibiotics in treatment of biofilm-associated infections, *American Society of Microbiology News*, **58**, 665–8.

Baross, J.A. and Morita, R.Y. (1978) Microbial life at low temperatures: ecological aspects, *Microbial Life in Extreme Environments* (ed. D.J. Kushner), pp. 9–71, Academic Press, London.

Bazin, M.J. and Prosser, J.I. (1992) Modelling microbial ecosystems, *Journal of Applied Bacteriology (Supplement)*, **73**, 89–95.

Booth, I.R., Cairney, J., Sutherland, L. and Higgins, C.F. (1988) Enteric bacteria and osmotic stress: an integrated homeostatic system, *Journal of Applied Bacteriology (Supplement)*, **65**, 35S–49S.

Carpentier, B. and Cerf, O. (1993) Biofilms and their consequences, with particular reference to hygiene in the food industry, *Journal of Applied Bacteriology*, **75**, 499–511.

Cavalier-Smith, T. (1987) Eukaryotes with no mitochondria, *Nature*, **326**, 332–3.

Chesbro, W., Evans, T. and Eifert, R. (1979) Very slow growth of *Escherichia coli*, *Journal of Bacteriology*, **139**, 625–38.

De Graeve, K.G., Grivet, J.P., Durand, M. *et al.* (1994) Competition between reductive acetogenesis and methanogenesis in the pig large-intestinal flora, *Journal of Applied Bacteriology*, **76**, 55–61.

Evans, H.C. (1988) Coevolution of entomogenous fungi and their insect hosts, *Coevolution of Fungi with Plants and Animals* (eds. K.A. Pirozynski and D.L. Hawksworth), pp. 149–71, Academic Press, London.

Fernandez-Linares, L., Faure, R., Bertrand, J.-C. and Gauthier, M. (1996) Ectoine as the predominant osmolyte in the marine bacterium *Marinobacter hydrocarbonoclasticus* grown on eicosane at high salinities, *Letters in Applied Microbiology*, **22**, 169–172.

Gerber, S. and Hader, D.-P. (1994) Effects of enhanced UV-B irradiation on the red coloured freshwater flagellate *Euglena sanguinea*, *FEMS Microbiology Ecology*, **13**, 177–84.

Goumelon, M., Cillard, J. and Pommepuy, M. (1994) Visible light damage to *Escherichia coli* in seawater: oxidative stress hypothesis, *Journal of Applied Bacteriology*, **77**, 105–12.

Gounot, A.-M. (1991) Bacterial life at low temperature: physiological aspects and biotechnological implications, *Journal of Applied Bacteriology*, **71**, 386–97.

Herbert, R.A. (1992) The application of microelectrodes in microbial ecology, *Journal of Applied Bacteriology (Supplement)*, **73**, 164–72.

Hicks, S.J. and Rowbury, R.J. (1987) Bacteriophage resistance of attached organisms as a factor in the survival of plasmid-bearing strains of *Escherichia coli*, *Letters in Applied Microbiology*, **4**, 129–32.

Kaprelyants, A.S., Gottschal, J.C. and Kell, D.B. (1993) Dormancy in non-sporulating bacteria, *FEMS Microbiology Reviews*, **104**, 271–86.

Keith, S.M., Russ, M.A., Macfarlane, G.T. and Herbert, R.A. (1987) The ecology and physiology of anaerobic bacteria isolated from Tay estuary sediments, *Proceedings of the Royal Society of Edinburgh*, **92B**, 323–33.

Kersters, I., Huys, G., van Duffel, H. *et al.* (1996) Survival potential of *Aeromonas hydrophila* in fresh water and nutrient poor waters in comparison with other bacteria, *Journal of Applied Bacteriology*, **80**, 266–76.

Lee, J.V. and West, A.A. (1991) Survival and growth of *Legionella* species in the environment, *Journal of Applied Bacteriology (Supplement)*, **70**, 121–30.

Lewis, D.L. and Gattie, D.K. (1990) Effects of cellular aggregation on the ecology of microorganisms, *American Society of Microbiology News*, **56**, 263–8.

Malloch, D.W. (1988) The evolution of mycorrhizae, *Canadian Journal of Plant Pathology*, **9**, 398–402.

Moran, M.A., Rutherford, L.T. and Hodson, R.E. (1995) Evidence for indigenous Streptomyces populations in a marine environment determined with a 16S rRNA probe, *Applied and Environmental Microbiology*, **61**, 3695–3700.

Mouget, J.-L., Pakhama, A., Lavoie, M.C. and de la Noue, J. (1995) Algal growth enhancement by bacteria: Is consumption of photosynthetic oxygen involved?, *FEMS Microbiology Ecology*, **18**, 35–44.

Moyer, C.L., Dobbs, F.C. and Kail, D.M. (1994)

Estimation of diversity and community structure through restriction fragment length polymaplian distribution analysis of bacterial 16S rRNA genes for a microbial mat at an active hydrothermal vent system, Loihi Seamount, Hawaii, *Applied and Environmental Microbiology*, **60**, 871–9.

Muller, R.H. and Babel, W. (1996) Measurement of growth at very low rates, an approach to study the energy requirements for the survival of *Alcaligenes eutrophus* JMP 134, *Applied and Environmental Microbiology*, **62**, 147–51.

Nielsen, T. and Eklund, N.G.A. (1995) Influence of solar ultraviolet radiation on photosynthesis and motility of marine phytoplankton, *FEMS Microbiology Ecology*, **18**, 281–8.

Nojoumi, S.A., Smith, D.G. and Rowbury, R.J. (1995) Tolerance to acid in pH 5.0 grown organisms of potentially pathogenic Gram-negative bacteria, *Letters in Applied Microbiology*, **21**, 359–63.

Oliver, J.D., Hite, F., McDougald, P. *et al.* (1995) Entry into, and resuscitation from the viable but non culturable state by *Vibrio vulnificus* in an estuarine environment, *Applied and Environmental Microbiology*, **61**, 2624–2630.

Ozkanka, R. and Flint, K.P. (1996) Alkaline phosphatase activity of *Escherichia coli* starved in sterile lake water, *Journal of Applied Bacteriology*, **80**, 252–8.

Pirozynski, K.A. and Hawksworth, D.L. (1988) Coevolution of fungi with plants and animals: Introduction and overview, *Coevolution of Fungi with Plants and Animals* (eds. Pirozynski, K.A. and Hawksworth, D.L.), pp. 1–29, Academic Press, London.

Rajendram, N., Matsuda, O., Urushigawa, Y. and Simidu, U. (1994) Characterization of microbial community structure in the surface sediment of Osaka Bay, Japan by phospholipid fatty acid analysis, *Applied and Environmental Microbiology*, **60**, 248–57.

Rehnstam, A.-S., Backman, S., Smith, D.C. *et al.* (1993) Blooms of sequence-specific culturable bacteria in the sea, *FEMS Microbiology Ecology*, **102**, 161–6.

Roszak, D.B. and Colwell, R.R. (1987) Survival strategies of bacteria in the natural environment, *Microbiological Reviews*, **51**, 365–79.

Russell, J.B. (1992) Another explanation for the toxicity of fermentation acids at low pH: anion accumulation versus uncoupling, *Journal of Applied Bacteriology*, **73**, 363–70.

Russell, N.J. (1990) Cold adaptation of micro-organisms, *Philosophical Transactions of the Royal Society of London B*, **326**, 595–611.

Russell, N.J. and Fukunaga, N. (1990) A comparison of thermal adaptation of membrane lipids in psychrophilic and thermophilic bacteria, *FEMS Microbiology Reviews*, **75**, 171–82.

Sutherland, L., Cairney, J., Elmore, M.J. *et al.* (1986). Osmotic regulation of transcription: Induction of the proU betaine transport gene is determined by the accumulation of intracellular potassium, *Journal of Bacteriology*, **168**, 805–814.

Woolfaardt, G.M., Lawrence, J.R., Robarts, R.D. *et al.* (1994) Multicellular organization in a degradative biofilm community, *Applied and Environmental Microbiology*, **60**, 434–6.

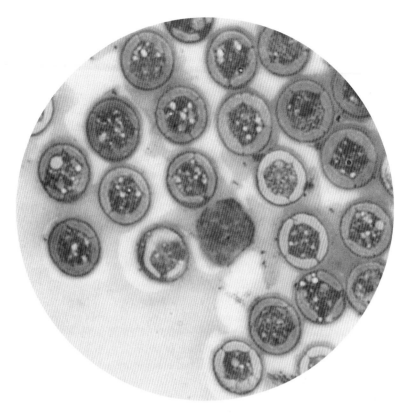

2 AQUATIC ENVIRONMENTS

In terms of physical size, the aqueous environment is the largest. The oceans represent the greatest area, but inland lakes and rivers are by no means insignificant. It has, in fact, been argued that the distinction between the oceans and freshwater environments is artificial and that all form part of the same continuum. In practice both oceans and freshwater systems contain a number of related, but distinct environments; conditions in the ocean depths, for example, are very different to those in the surface layers, which again differ from those in estuaries. Equally conditions in a eutrophic inland lake are likely to be very different to those in a low-nutrient mountain stream. With the exception of specific circumstances that permit rapid microbial growth, however, aquatic environments tend to be severe and successful micro-organisms deploy a variety of survival strategies. Strategies are also required to deal with competition from and co-operation with other micro-organisms. Despite these factors many aquatic environments support a wide range of micro-organisms, from phage through eubacteria and cyanobacteria to microfungi, protozoa and eukaryotic algae.

MICROFLORA OF THE DIFFERENT AQUATIC ENVIRONMENTS

Introduction

Aquatic environments range from the pristine stream emerging from the mountain top to the chimneys of hydrothermal vents in the deepest parts of the ocean. At the same time, ecosystem habitats may range in size from microniches on small pieces of particulate matter to stratified lakes and microbial mat communities. It has, however, been argued that, with the exception of specialized micro-organisms, the ecology of marine and freshwater micro-organisms is similar. The extent to which this is true is debatable, but some microbiologists consider that micro-organisms of marine and freshwater origin coexist in estuarine and inshore environments where fresh water and sea water mix. Further evidence is required for a full understanding of the situation. It is established, however, that there are many common features of life in photic aquatic communities where algae and cyanobacteria are of much greater importance in primary production than in most terrestrial environments.

The importance of predation in controlling the bacterial population has long been recognized in both aquatic environments, where hetero-trophic nano-flagellates are important consumers of bacteria and play a major role in the regeneration of nutrients. There is considerable diversity among the hetero-trophic nanoflagellates, and their relationship with the prey population can vary considerably (*Feature 3*).

A role for predatory bacteria in limiting the size of the overall bacterial population has been discussed, but is generally thought to be very limited. The parasitic bacterium *Bdellovibrio bacteriovorus* (**104–106**) is present in seawater and freshwater, but its overall importance is probably limited. *Bdellovibrio bacteriovorus* may, however, be of much greater importance in restricted niches where the prey density is high. The gliding aquatic myxobacter, *Lysobacter* (**107, 108**) is algicidal and may limit the extent of blooms.

The importance of bacteriophage activity (**15–17**) in aquatic environments of all types has only recently been appreciated. The properties of the phage population of aquatic environments are poorly understood but it is considered that in the oceans and probably other waters, viruses are the most numerous micro-organism. Bacteriophages are of considerable ecological importance, not only for limiting the numbers of other micro-organisms but also because nutrients released after lysis are available for direct utilization only by other micro-organisms. In this way nutrient movement upwards through the food chain is short-circuited. It has been argued that bacteriophage activity, while potentially lethal for individual micro-organisms, actually benefits the microbial community as a whole by maintaining nutrient availability. Equally, it has been demonstrated, using fluorescently labelled probes, that viruses can, although not invariably, control the structure of marine microbial communities at levels below one infective unit ml^{-1} (Hennes *et al.*, 1995).

It is also recognized that micro-organisms in aquatic environments are subject to starvation and to long periods at low temperatures. There is, however, a matter of magnitude, e.g. bacteria carried from ocean surface layers to the depths by water upheavals may have to survive cold, starvation and high pressure for many years before returning to more favourable conditions. In contrast, deprivation in freshwater environments may be regarded as minimal and short-lasting.

Feature 3. Feeding strategies

Heterotrophic nanoflagellates are a complex group of organisms that are able to coexist by employing different survival strategies and occupying different ecological niches. This may be demonstrated by the differing bacterial consumption patterns of starved cells of a filter feeder and an interception feeder. The filter feeder *Pteridomonas danica* continuously consumed four to five bacteria per hour over an eight hour period, irrespective of bacterial density above 5×10^5 ml^{-1}. Catching continued at lower bacterial densities. In contrast, the interception feeder *Paraphymonas imperforata* (which has only half the biomass of *Pteridomonas danica*) ingested ten bacteria per hour at high bacterial densities but reduced feeding rate with reducing densities. Feeding ceased at a bacterial density of 1×10^6 ml^{-1}.
Data from Zubkov and Sleigh, 1995.

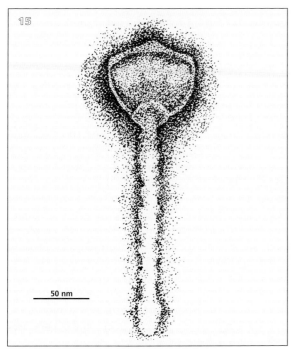

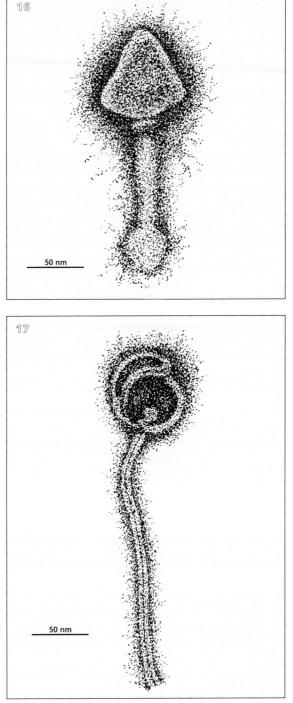

15–17 Morphology of typical viruses isolated from an aquatic (lake) environment.

Blooms

Blooms are a feature of both seawater and freshwater environments. Algal and cyanobacterial blooms are well recognized, but less attention has been paid to bacterial blooms.

Algal and cyanobacterial blooms

Algal blooms occur as a consequence of eutrophication, an increase in nutrient load, of bodies of both fresh and salt water. Algal blooms appear to be increasing in frequency, at least partially as a consequence of man's activities, especially increased levels of phosphorus and nitrogen. Algal blooms are natural occurrences and the spring phytoplankton bloom in temperate and arctic seas is a well-known phenomenon. Blooms are predictable consequences of hydrological changes which increase the nutrient content of upper layers of water. These include run-off from land, water column turnover and up-welling, bringing deep, nutrient-rich water to the surface. The spring bloom results from the onset of stratification and a high nutrient level due to vertical mixing, coinciding with increased day length and quantity of solar radiation.

Under favourable conditions, the growth of phytoplankton is exponential. Algal biomass and bloom formation may be controlled by herbivore grazing but response is slow, permitting numbers to rise rapidly to $c.10^8$ ml^{-1}. Numbers of algal phages increase rapidly with increasing cell numbers and metabolic activity. The role in preventing bloom formation is not known but on at least one reported occasion a bloom has been stopped by phage. Species succession occurs as conditions change but competition means that, at their peak, blooms are usually monospecific, *Phaeocystis* for example being most common in spring blooms in coastal North Sea zones. Blooms are often recognized visually by discoloration of the sea which may be green, brown, or red depending on the algae present. White blooms are due to the calcareous skeletal elements of coccolithophores, such as *Emiliania huxleyi*, while 'red tides' are caused by dinoflagellates.

The high levels of organic material produced by blooms can have beneficial effects further up the food chain though, in many cases, production of organic matter is so great, that remineralization by bacteria consumes very large quantities of dissolved oxygen. This, together with oxygen consumption by the night-time respiration of algae, leads to anoxia and mass mortality amongst fish and invertebrates. Fish may also be killed by toxins and by H_2S production. The effect of blooms is often particularly marked in closed systems such as lakes; long-term effects on ecosystems often result. Cyanobacterial blooms reduce light which may ultimately lead to the disappearance of larger aquatic plants and visually hunting piscivorous fish. Filter-feeding planktonivorous fish become dominant leading to the virtual disappearance of crustacean zooplankton, reducing the grazing pressure on algae.

From the viewpoint of man, blooms are undesirable for a number of reasons. They are foul smelling and their presence results in a loss of recreational value in both coastal and inland waters. A number of genera produce toxins which may either directly affect man and animals entering the water, or indirectly through consumption of shellfish that have acquired algal toxins. Not all algae involved in bloom formation are toxigenic but toxins may be involved in antipredator strategies and toxigenic species are favoured under conditions supporting bloom formation. In recent years some toxigenic algae have spread geographically. *Gymnodinium catenatum*, for example, has spread from southern Californian waters to southern Europe, Japan and Australia. Also, the previously non-toxigenic genus *Nitzschia pungens* is now recognized as producing a toxin – domoic acid – and is associated with amnesic shellfish poisoning.

Bacterial blooms

Bacterial blooms are less visible than algal or cyanobacterial blooms and have a less detrimental effect on the environment. Bacterial blooms occur in near-shore and freshwater environments under similar conditions to algal and cyanobacterial blooms.

The role of aquatic micro-organisms in the formation of carbonate minerals

A number of micro-organisms, of which cyanobacteria are of considerable importance, are involved in the formation of carbonate minerals. Initial work suggested that this phenomenon was confined to marine systems but cyanobacteria are now known to be essential mediators of calcite formation in freshwater environments (Schulzte-Lam and Beveridge, 1994).

In water containing abundant Ca^{2+} ions and dissolved CO_2 – primarily as HCO_3^- in alkaline waters – cyanobacteria provide nucleation sites for mineral formation to commence. Utilization of HCO_3^- leads to the release of OH^- ions and a rise in pH value in the microenvironment surrounding the cell. The formation of carbonate minerals is highly pH dependent and the degree of alkalization determines the mineral phase developing. Carbonate phases are deposited at pH values in excess of 8.0 while other mineral phases are formed at lower pH values.

Studies of a meromictic lake of high Ca^{2+} and SO_4^{2-} content showed gypsum ($CaSO_4 . 2H_2O$) to be initially formed on cells of *Synechococcus*. As cellular activity increased, however, the dominant phase became calcite. Calcitic microbial reefs were present on the lake shore and marl sediments deposited.

It is now known that biogenesis of other carbonate minerals, which were previously considered to be of non-biological origin, is possible (Schulzte-Lam and Beveridge, 1994). These include strontianite ($SrCO_3$) and magnesite ($MgCO_3$).

THE MARINE ENVIRONMENT

Introduction

Conditions within the marine environment as a whole vary widely and it is necessary to discuss the various major parts separately. It is customary to categorize the various types of micro-organism biologically and by size. Thus the bacterio- or pico-plankton are in the size range 0.2–2.0 μm, much of this population being ultramicrobacteria. The phytoplankton (plants) is subdivided into nano-plankton (2.0–20 μm) and microplankton (20–200 μm). Microscopic animals are zooplankton and bacteriophages femtoplankton.

Many of the bacteria from marine environments are characterized by an obligate requirement for Na^+ and other ions in seawater. Some of these are closely related to freshwater bacteria, which have no such requirement or a requirement at a significantly lower level. Chitinolytic activity is extremely common amongst marine bacteria as a consequence of the vast supplies of chitin available from the bodies of marine arthropod exoskeletons. Agarolytic activity is also widespread, especially amongst marine cytophages.

The dissolved organic matter (DOM) present in seawater is one of the largest pools of organic matter in the biosphere. Dissolved organic matter is a major substrate for heterotrophic bacteria and an intermediate in the carbon flux from phyto-plankton and zooplankton to bacteria. Only a discrete fraction is available as substrate for heterotrophs, the remainder being highly recalcitrant with a turnover time which may be as long as several thousand years. Dissolved humic matter (DHM) accounts for most of this recalcitrant material. This material has a similar $^{13}C:^{12}C$ ratio to phytoplankton and is distinct from humic material derived from terrestrial plants. It therefore appears that a substantial quantity of recalcitrant DOM, including DHM, is a product of primary production (Tranvik, 1993). Dissolved humic matter probably comprises the recalcitrant fraction of the DOM together with complex, recalcitrant matter produced from the labile fraction of DOM, possibly as a consequence of sequential assimilation and excretion by several different micro-organisms.

The ocean

As a consequence of its great depths, the ocean consists of several broad groups of habitat of differing physicochemical characteristics (**18**). These may be grouped into the sea surface microlayer, inhabitants of which are often categorized as the neuston. Surface waters above the thermocline, which are illuminated and heated by solar radiation, comprise the epipelagic habitat, while immediately below the thermocline is the mesopelagic habitat. This merges with the bathypelagic at a depth of c.2,000 m. The average depth of the ocean bed is 3,800 m, below which the habitat is described as abyssopelagic. At an even greater depth are hadal habitats, which extend below 11,000 m.

The sea surface microlayer

The sea surface microlayer is a thin film of liquid at the interface of the sea and the atmosphere, which is of considerable importance in exchanges between sea and atmosphere. The film is oily, consisting primarily of lipids derived – in the absence of pollution – from external sources, from marine organisms and natural hydrocarbon seeps.

18 Classification of habitats in relation to the topography of the ocean. All habitats in the water column are pelagic while sea bed habitats are benthic. Both pelagic and benthic habitats in the continental shelf region are included in the neritic habitat.

```
18
---------- High water ----------------------
Littoral
---------- Low water ------------------------
Epipelagic (photic)
---------- Thermocline ----------------------
Mesopelagic
---------- 2,000 m --------------------------
Bathypelagic
---------- 3,800 m --------------------------
Abyssopelagic
---------- Oceanic trenches -----------------
Hadal
---------- 11,000 m -------------------------
```

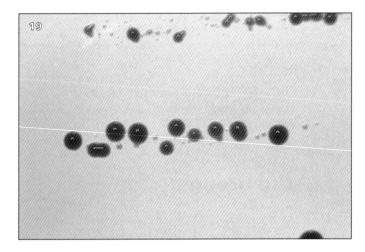

19 A red-pigmented, marine psychrophile. The isolate appeared to resemble '*Vibrio psychroerythrus*' which shares some properties with *Photobacterium*. DNA homology studies, however, suggest no relationship with either *Photobacterium* or *Vibrio*. (Source: seawater held at 3°C (37.4°F) for 4 months; Medium: yeast extract–artificial seawater medium; Incubation: *c.*12°C (53.5°F), 10 days.)

Bacteria are transported to the surface layer with rising gas bubbles, those with hydrophobic properties preferentially accumulating there. Conditions at the surface layer are harsh, organisms facing a high level of visible and UV radiation and a high concentration of pollutants. The fats and oils forming the layer, however, provide ideal substrates for growth, while exposure to pollutants leads to rapid development of consortia capable of degrading them. Micro-organisms present at the surface layer are highly resistant to UV irradiation and to chemical abuse. Pigmentation may also offer protection from UV irradiation (**19**).

Resistance to pollutants and degradative abilities mean that surface layer bacteria are of considerable importance in bioremediation. In the case of oil spillage, surfactants are sprayed on to the surface to break the oil into droplets and increase the surface area available to bacteria, but this practice can have undesirable consequences. In all cases provision of additional nitrogen and phosphorus, which are often limiting, enhances microbial activity and metabolism of pollutants.

Epipelagic habitats

Light penetrates into epipelagic habitats and so photosynthesis can occur. In the open ocean the epipelagic zone may be up to 150 m deep compared with only 15 m in the turbid waters of the neritic zone. Under normal conditions, the epipelagic zone is separated from the deep ocean by the thermocline but is well mixed by convection currents arising through solar heating. The water is oxic although the O_2 content may be significantly reduced by bacterial metabolism, especially after algal blooms.

Temperatures range from –1°C (30°F) in polar waters to *c.*35°C (95°F) in the tropics.

Primary production is through photosynthesis by cyanobacteria and prochloroplasts. *Synechococcus* is well adapted to its habitat at the bottom of the epipelagic zone, its photosynthetic pigments adsorbing light strongly at 400–500 nm, the wavelength of light penetrating seawater. The organism is also of small, coccoid morphology, which minimizes sinking. A variety of aerobic marine bacteria have been isolated that synthesize bacteriochlorophyll *a*. It has been suggested that such organisms comprise 1–2% of the eubacterial population. The significance of this synthesis is not known in all cases but in the most widely studied of these bacteria, the strictly aerobic *Erythrobacter*, it appears that almost 60% of ATP synthesis is light driven. This may permit the more efficient utilization of organic compounds and enhance growth in conditions of carbon limitation. Anoxygenic phototrophic bacteria are not normally of significance in open oceans except in stratified marine basins such as the Black Sea. In the Black Sea and similar seas, where anoxic, sulphide-rich water occurs, a population of the green sulphur bacteria *Chlorobium* develops beneath the chemocline between the O_2- and H_2S-containing layers, at a depth of 60–110 m (cf. stratified lakes, page 74).

The concentration of nutrients of plant origin, nitrite, ammonia, urea, phosphate and trace minerals, is often very low in the epipelagic zone. Organic matter is recycled with very little input from other environments, although substrates are available for heterotrophic bacteria as detritus and organic aggregates settle through the water column. Exoenzymes are of considerable importance in the degradation of complex molecules and in some cases several different bacteria are involved.

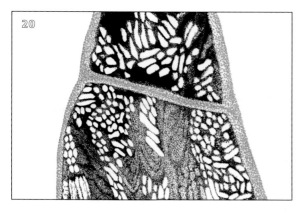

20 Structure of gas-vacuoles (× 26,000). Vacuoles are compound organelles consisting of a variable number of individual, spindle-shaped gas vesicles. Each vesicle is bounded by a 2 nm protein layer and banded by regular rows of sub-units.

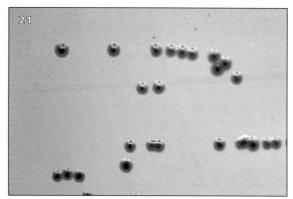

21 A species of *Alteromonas*. The isolate produces a weakly diffusible brown pigment but differs phenotypically from the brown-pigmented species of *Alteromonas*, *A. hanedai*. [Source: seawater; Medium: yeast extract–artificial seawater agar (Baumann *et al.*, 1984); Incubation: c.18°C (64.5°F), 7 days.]

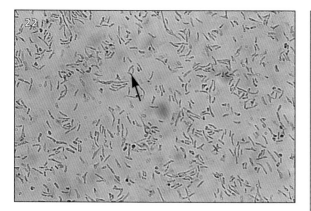

22 A species of *Alteromonas*. Irregular cell shapes are common amongst marine species of Gram-negative bacteria and are usually associated with senescence or adverse growth conditions. Swellings (arrowed) are not endospores but are due to cell wall weakness, possibly as a consequence of unfavourable ionic conditions. (Source: as for **21**; Microscopy: Gram stain, × 1,200.)

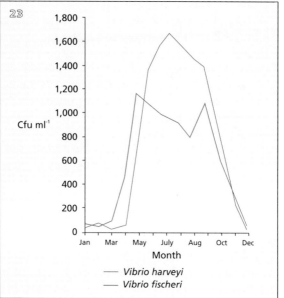

23 Seasonal distribution of *V. harveyi* and *V. fischeri* in Pacific waters off southern California. Different patterns have been reported elsewhere but, in general, there is a good correlation with water temperature.

Nutrients recycled by biodegradative processes and by zooplankton digestion are immediately assimilated by bacteria and phytoplankton. A variety of heterotrophic bacteria are known to be common inhabitants of the epipelagic zone. These are generally small, Gram-negative rods, some of which are gas vacuolated (**20**); gas vacuolation is assumed to be beneficial in maintaining bacteria at the optimal depth. Genera present in significant numbers include *Alteromonas* (**21**, **22**), *Marinomonas*, *Photobacterium* and marine *Vibrio* species. Distribution of these bacteria is variable according to other factors, especially water temperature. Species of *Vibrio* show a considerable variation in distribution according to season (**23**). Cells enter the viable, non-recoverable state when water temperatures fall below c.15°C (59°F) but recover rapidly when the temperature is c.21°C (70°F) (Oliver *et al.*, 1995). This pattern can vary with geographical location. Epipelagic bacteria may also be present on the surface of fish as saprophytes while *Photobacterium* and some species of *Vibrio* form bioluminescent symbioses (see pages 43–45).

Feature 4. The Great Magic Trick

Cyanobacteria, such as *Trichodesmium*, *Nostoc*, *Nodularia* and others, are of very significant ecological importance in oligotrophic waters as a consequence of their ability to fix both carbon and nitrogen. The 'Great Magic Trick' for these micro-organisms is to manage both oxygenic photosynthesis and nitrogen fixation, the latter being an extremely oxygen-sensitive process. The basic problem is to protect the highly oxygen-sensitive enzyme nitrogenase from both endogenously evolved oxygen and that in the environment (Fay, 1992).

Different cyanobacteria have adopted different management strategies to the problem of both oxygenic photosynthesis and nitrogen fixation. *Plectonema*, for example, has adopted a very simple strategy of fixing nitrogen only under conditions of low oxygen (micro-oxic) tension. A further simple strategy is that of cyanobacteria such as *Oscillatoria limnosa* which fix nitrogen only during the dark, non-photosynthetic hours of the night. The most sophisticated approach appears to lie with those cyanobacteria that have evolved heterocysts – specialized anaerobic cells lacking oxygen-evolving mechanisms and which have additional cell wall layers to minimize oxygen penetration. Cyanobacteria of this type include *Anabaena*, *Nostoc* (**25**), *Nodularia* (**26**, **27**) and *Aphanizomenon* (**28**).

Trichodesmium, however, is the greatest magician of all! This organism has the ability to fix nitrogen during the day while lacking the specialized oxygen-protective heterocysts. It has been extensively studied at the University of Stockholm using immunogold/transmission electron microscopy and immunofluorescence/light microscopy. Examination of three species of *Trichodesmium* (**29–31**) showed nitrogenase to be restricted to a limited (*c.*14%) subset of cells randomly distributed throughout the colony. The frequency of nitrogenase-containing cells varied diurnally, being highest during the day, the period of active nitrogen fixation and falling overnight to approach zero just before dawn (**32–36**, see page 38). It appears that colonies of *Trichodesmium* contain specialist nitrogen-fixing cells or whole filaments. Segregation of nitrogenase to specialist cell types may be seen as analogous to heterocyst production and the pattern of nitrogen fixation (only during daytime and with a peak at noon) is the same as heterocystous cyanobacteria. Although a number of theories have been advanced it is not known how the nitrogenase is protected from oxygen.

Data based on Fredriksson and Bergman, 1995.

With the exception of equatorial and subarctic areas of the Pacific ocean, open epipelagic waters are nitrogen deficient. Despite this, N_2 fixation by either cyanobacteria or N_2-fixing organotrophs is unusual. In these circumstances the non-heterocystous cyanobacterium *Trichodesmium* (**24**), which is capable of concurrent N_2 and CO_2 fixation, is of major importance in primary nitrogen and carbon cycling. *Trichodesmium* is a colony-forming cyanobacterium occurring over very large areas of tropical and subtropical oceans (*Feature 4*). It is stimulated by iron, the loss of which through sedimentation is an acute and chronic problem. The

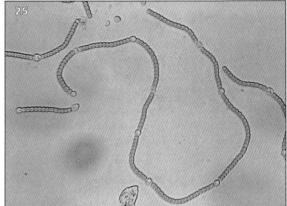

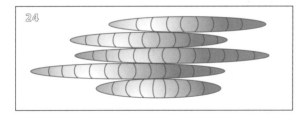

24 Diagrammatic representation of cells of *Trichodesmium*, a filamentous cyanobacterium. Filaments often develop as parallel bundles of 'trichomes'.

25 Heterocystous cyanobacteria. *Nostoc* is involved in symbiotic relationships, the strain illustrated having been isolated from *Gunnera monoika*. *Nodularia* has a regular pattern of heterocysts throughout the filament.

26 Heterocystous cyanobacteria. A strain isolated from the Baltic Sea.

27 Heterocystous cyanobacteria. Autofluorescence; there is no fluorescence from the heterocysts. In contrast to *Nodularia*, *Aphanizomenon* produces few heterocysts (one or less per filament).

28 Heterocystous cyanobacteria. A Baltic Sea isolate is illustrated, showing autofluorescence even from heterocysts. **25–28** reproduced by courtesy of Ulla Rasmussen, Department of Botany, University of Stockholm, Sweden.

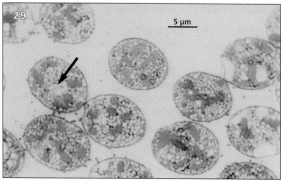

29 One of three species of *Trichodesmium* representing the major fusiform species in the Caribbean. Colonies, which have been cross-sectioned and stained with toluidine-blue, were classified primarily according to cell size and gas vacuole (arrowed) arrangement. *Trichodesmium thiebautii* has the largest cell diameter of the three species and a number of gas vacuoles scattered over the cells.

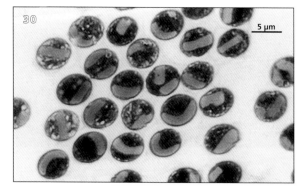

30 The second of the three species, *Trichodesmium tenui*, produces a large number of filaments in each colony and is further characterised by one, or a few, gas vacuoles per cell.

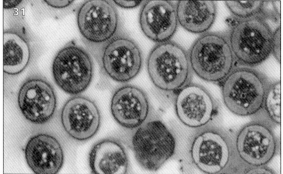

31 The third of the three species, *Trichodesmium erythraeum*, is markedly less common in the Caribbean than *Tr. thiebautii* and *Tr. tenui*. The organism is characterised by the distinct cylindrical gas vacuole 'tube' in the cell periphery. **29–31** reproduced with permission from Fredriksson and Bergman, 1995, © 1995 The Society for General Microbiology.

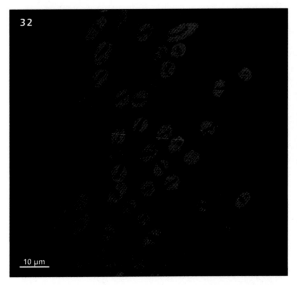

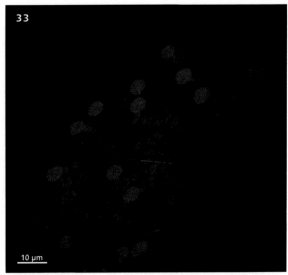

32 The illustrations **32–36** show how nitrogenase activity in *Tr. tenui* colonies is revealed by immunofluorescence/light microscopy. Cells were labelled for the Fe protein of nitrogenase and clearly show the random distribution within the colony. The series demonstrates the variation in intensity of the label (fluorescence) and thus nitrogenase activity at different time points. This is at 06.10.

33 At 08.00.

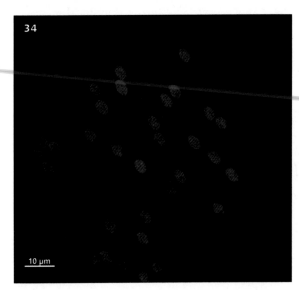

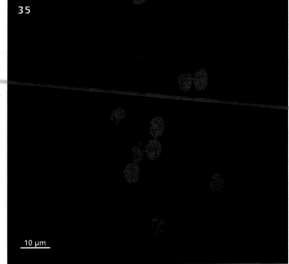

34 At 12.00 (time of maximum nitrogen fixing activity).

35 At 14.00.

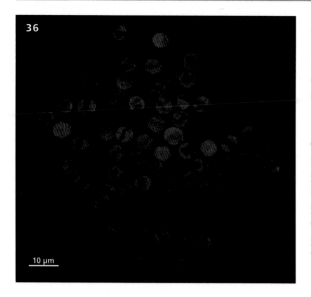

36

10 µm

36 At 19.30. **32–36** reproduced with permission from Fredriksson, C. and Bergman, B., 1995, © 1995 The Society for General Microbiology.

epipelagic zone is usually, therefore, iron deficient depending on input from the atmosphere, terrestrial discharge and upwelling for new iron.

Considerable quantities of ammonia are produced by the degradation of organic matter. In addition to that assimilated by bacteria and phytoplankton, a significant quantity is oxidized to NO_2 and, subsequently, to NO_3 by nitrifying bacteria. These bacteria are also responsible for biogenesis of the 'greenhouse gas' N_2O. Biogenesis is prevalent at low oxygen levels and is probably mediated by a nitrate reductase but this source is of minimal significance compared with fertilizer nitrification.

Urea is an important source of nitrogen for marine phytoplankton and may be used in preference to nitrate and nitrite. The presence of urea can partly be accounted for by excretion from zooplankton but other sources are obviously involved, suggested sources including guano, larval fish, sharks, terrestrial outflow, microbial degradation of arginine and purines and flux from sediments. Of these, it is thought that bacteria in sediments may be an important source of urea in

the water column (Pedersen *et al.*, 1993).

Cyanobacteria in epipelagic and other marine habitats can play a significant role in climatic change. Dimethylsulphopropionate (DMSP) is an osmolyte synthesized by coccolithophores such as *Emiliania huxleyii*. Senescence and zooplankton feeding leads to the release of DMSP, providing a characteristically marine substrate for bacteria. One pathway for metabolism of DMSP results in the production of dimethyl sulphide (DMS) which is volatile and enters the atmosphere. Bacterial production of DMS from DMSP accounts for 90% of natural sulphur entering the atmosphere and produces sulphuric and methanesulphonic acids, which promote cloud formation.

Midwater habitats

Temperatures in midwater habitats below the thermocline are normally 0–5°C (32–41°F). Insufficient light penetrates to support photosynthesis and so the habitat is permanently dark. Organic matter is degraded by heterotrophic microorganisms, with regeneration of inorganic nutrients for phytoplankton and bacteria in the photic epipelagic zone. There is only a very limited return of nutrients by vertical diffusion and most occurs during wind-driven upwelling. Nitrification occurs but the significance of this process has been doubted, though the importance of ammonia-oxidizing bacteria may have been underestimated; ratios as high as $10^4:1$ of *Nitrosomonas* to *Nitrococcus* have been reported. Two other genera, *Nitrosolobus* and *Nitrosospira*, are present in smaller numbers (Sinigalliano *et al.*, 1995). In some cases, methane is available from seeps. Methophils, however, are uncommon in midwater habitats and methane may be oxidized by ammonia-oxidizing bacteria.

Oxygen consumption by heterotrophic bacteria often proceeds at a sufficiently high rate to permit the development of anoxic or micro-oxic zones. These are particularly common in areas, such as the northwest Pacific, where water has been out of contact with the atmosphere for long periods. Nitrate respiration and denitrification, with concomitant oxidation of organic matter, is common if NO_3 is available.

Organic particles and aggregates of various size provide sites for attachment of bacteria and there is also attachment to inorganic particles, where nutrients tend to concentrate. Large particles, comprising mucoid aggregates, are commonly known as 'marine snow' and are of considerable importance in the transport of organic matter to deeper regions. Diffusion within the particles is limited and permits the establishment of stable pH and oxygen gradients, allowing for the establishment of an anaerobic community (Rath and Herndl, 1994).

Unattached bacteria face less favourable nutritional conditions than those attached to particles. Adaptation occurs through size reduction and metabolic adjustment though this adaptation should not be confused with a survival state, but may be considered nutritionally 'opportunistic' (Taylor, 1992), response being very rapid when a higher level of nutrients becomes available.

Deep-sea habitats

These habitats are generally inhospitable environments, being cold, energy-poor and under extreme pressure. Only about 1% of primary production in surface waters reaches the depths, while pressure on the ocean bed averages 400 atmospheres and is as high as 1,100 atmospheres in the deepest trenches. Metabolism of bacteria in the water column and in sediments is slow. This, however, is an adaptation to the energy-poor nutrient status and not to the deep-sea status *per se*. In copiotrophic environments, such as the guts of deep-sea amphipods, or localized concentrations of detritus, metabolism can be relatively rapid.

Most bacterial activity in the water column is attributed to aerobes but in some marine areas of Antarctica, and possibly elsewhere, anaerobes play a substantial role in the metabolism of organic matter. This is probably a consequence of the high organic input. Truly psychrophilic anaerobes, however, are rare in any environment and most Antarctic isolates are psychrotrophic rather than psychrophilic. Essentially anaerobic metabolic activity, such as sulphate reduction, does occur in permanently cold sediments though the lowest reported optimum temperature for sulphate reduction is 21°C (70°F), in contrast with the *in situ* temperature of −1 to 1°C (30.2 to 33.8°F) (Isaksen and Jorgensen, 1996). This metabolic behaviour is consistent with the general temperature relationships of micro-organisms in cold environments but suggests a very low rate of activity. It is sometimes felt that the current paradigm describing the behaviour of micro-organisms at very low temperatures is insufficient in explaining what actually occurs in cold environments.

Where isolation and identification of deep-sea bacteria has been possible, most belong to the same genera as those in midwater and epipelagic habitats. Most show a high level of adaptation to the environment and are both psychrophilic or psychrotrophic, and barophilic. An extreme example was represented by a bacterial isolate from 10,500 m deep in the Mariana Trench, which was unable to grow at pressures less than 350 atmospheres and which died at atmospheric pressure. High pressures, like low temperatures, tend to solidify membranes, adaptation to both conditions requiring a considerable increase in the unsaturated fatty acid content of membranes. Bacteria from upper layers, which are occasionally carried to the depths by hydrodynamic forces, are unable to grow but can survive until carried back to more favourable levels by upwelling; sometimes many years later.

Hydrothermal vents provide abundant energy sources (H_2S, FeS, S, CH_4, NH_3) for bacteria but not other organisms. These are derived through the seepage of seawater at tectonic spreading centres and subsequent contact with magma (37). Thermal energy from radioactive reactions in the Earth's core provides thermal energy to drive subsequent reactions. Overall reactions involve the reduction of SO_4^{2-} to H_2S which reacts in turn with metals in hydrothermal fluids to form FeS, MnS, and the like, which are oxidized to elemental sulphur when hydrothermal fluids mix with oxygen-containing seawater. Carbon dioxide is reduced to CH_4 and H_2 and NH_3 are formed.

Temperatures at vent sites range from 5°C (41°F) at 'cold seeps' to 110°C (230°F). 'Black smokers' occur where hydrothermal fluid, leaving chimneys at 350–400°C (662–752°F), mixes directly with seawater. Oxidation of FeS and Mn^{2+} occurs forming $FeO(OH)$ and MnO_2 which subsequently sediment. At other sites seawater mixes with hydrothermal fluid before emission forming 'white smokers' which contain elemental sulphur.

Vents and their environs support populations of chemolithotrophic bacteria, most of which are also autotrophic. These include aerobic chemoautotrophs, sulphate reducers, methanogens and methophils. Oxic waters in vent habitats contain mats of sulphur-oxidizing bacteria including *Beggiatoa* (38–40), *Thiothrix* and *Thiovulum* (as well as unidentified strains) which develop as mats on

37 Diagrammatic representation of hydrothermal vents.

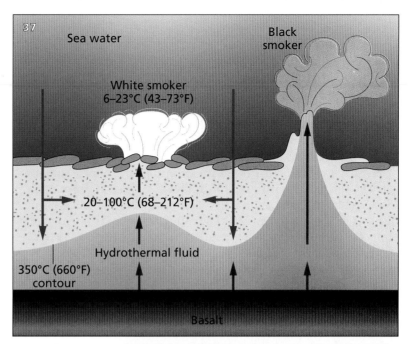

38 Phase contrast photomicrograph of cells resembling *Beggiatoa* isolated from the surface of hydrothermal sediment in a relatively shallow submarine biotope near Kodakara-Jima Island in Japan.

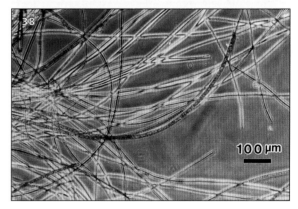

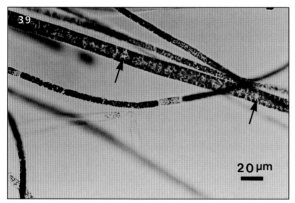

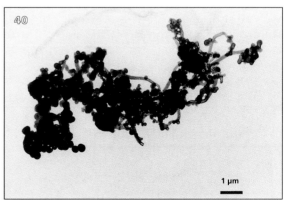

39 As for **38** but a bright field photograph, the arrows indicating intracellular sulphur particles. These bacteria are hyperthermophilic and have an obligate dependence on elemental sulphur.

40 Transmission electron micrograph of the *Beggiatoa*-like organism from a hydrothermal sediment. The organism has been negatively stained using uranyl acetate and was grown in a medium supplemented with 0.2% yeast extract and 1.0% elemental sulphur. **38–40** reproduced with permission from Hoaki (1995), © 1995 American Society for Microbiology.

sediments and even in the chimneys themselves. Growth is sufficient to support invertebrate food chains including filter feeding by clams and mussels, grazing by fish and symbioses (see below). Such chains ultimately depend on O_2 from photosynthesis for aerobic chemosynthesis. Chemosynthesis can, however, proceed independently of photosynthesis while combined nitrogen is supplied by free-living nitrogen fixers.

Hyperthermophilic environments exist in black smokers at temperatures of 80–110°C (176–230°F). Black smoker bacteria are usually strict anaerobes due to the low solubility of O_2 in water at high temperatures. Thermophilic, aerobic, non-endospore-forming, heterotrophic bacteria have, however, now been isolated from hyperthermophilic environments in deep-sea vents (Marteinsson et al., 1995). Bacteria from such environments exhibit considerable adaptation and cannot usually grow at temperatures below 60°C (140°F) although survival in cold, oxic water is possible for at least a year. Most strains are sulphidogenic, oxidizing H_2 with sulphur, sulphate and CO_2 and fixing CO_2. Heterotrophic growth may also be possible. Major metabolic reactions are summarized in **41**.

Animals and plants are unable to grow at temperatures above c.40°C (104°F) and are absent from the actual sites of hydrothermal vents. A very steep thermal gradient exists at vent sites, however, and animals are able to live in close proximity. Symbioses have developed between these deep-sea animals and chemolithotrophic bacteria living in the tissues. The first symbioses studied involved the utilization by bacteria of reduced energy sources, such as H_2S, present in hydrothermal fluids for autotrophic CO_2 formation. The hosts – tube worms, clams and mussels – benefit through an internal source of nutrients while the bacterial

symbionts receive inorganic substrates for chemosynthesis. At a later date, a symbiotic relationship was discovered between deep-sea invertebrates dwelling in the region of cold, methane seeps and methophilic bacteria. This symbiosis was initially thought to be restricted to cold seeps. Investigations at the 'Snakepit', an active hydrothermal site in the mid-Atlantic ridge, however, showed methophilic symbioses between bacteria and macrofauna rather than sulphur-oxidizing, autotrophic symbioses (Cavanaugh et al., 1992). Methophilic symbioses involve a simple, short and efficient flow of carbon from methane to the bacterium and from the bacterium to the animal. This may have biotechnological potential (Taylor, 1992).

Neritic and littoral habitats

Neritic (near-shore) and littoral (intertidal) habitats differ in a number of ways from oceanic habitats. Littoral habitats, in particular, are stressful environments for both marine and non-marine microorganisms. Despite this, turnover of organic compounds is much more intense and about 90% of bacterial degradation occurs in the benthic zone of the continental shelf region, which occupies only 10% of the total marine environment. Nutrients are available at a higher level than in ocean habitats and a wider range of organic substrates are present. These are metabolized by heterotrophic bacteria which are also essential for degradation of pesticides and other pollutants.

High light intensities lead to the extensive development of benthic plant communities, including seaweeds, marshgrasses and seagrasses. In tropical and subtropical climates, mangroves and coral reefs are also of importance. Primary production by plants is at a high level, enhanced by rapid regeneration of inorganic nutrients and input of nutrients from terrestrial sources. Bacteria are heavily involved in biogeochemical cycles: carbon, nitrogen, sulphur, phosphorus and trace metals. Combined nitrogen is recycled by the degradative activities of heterotrophic bacteria. Cyanobacteria develop very extensive mat communities which are of major importance both in nitrogen fixation and photosynthesis. A highly diverse range of prokaryotic micro-organisms in mats are capable of nitrogen fixation and the distribution of nitrogenase genes indicates that the ability to fix nitrogen cannot be predicted on the basis of taxonomy. A

41

Chemolithotrophic

$$H_2 + S° = H_2S$$

$$4H_2 + SO_4^{2-} = S^{2-} + 4H_2O$$

$$CO_2 + SO_4^{2-} = CH_4 + 2H_2O$$

Heterotrophic

$$CH_2O + 2S° + H_2O = CO_2 + 2H_2S$$

$$2CH_2O + SO_4^{2-} = S^{2-} + 2CO_2 + 2H_2O$$

41 Metabolism of hyperthermophiles.

study of nitrogen fixation in mats off the coast of North Carolina showed a seasonal shift, cyanobacteria dominating nitrogen fixation in summer and heterotrophic bacteria dominating in winter (Zehr *et al.*, 1995). Cyanobacteria on the leaves of seagrass also fix significant quantities of N_2 while *Azotobacter* forms a nitrogen-fixing endosymbiosis with the green seaweed *Codium* sp. Sulphate-reducing bacteria are also involved in N_2 fixation in sediments.

Genera of micro-organisms present in the water column of the ocean are also present in neritic and littoral environments. In some cases a wider range of species is present, although significant temperature differences between summer and winter may be reflected in considerable seasonal variation. *Vibrio* is more prevalent than in open oceans and species, including the human pathogens *V. cholerae* and *V. parahaemolyticus* (**42**). Both enter a viable, nonculturable stage during cold conditions, although survival of *V. cholerae* may also be linked with its association with chitin particles. *V. parahaemolyticus* has been described as a 'marine enteric organism' and is often associated with shellfish such as clams. This has public health implications as such products are often eaten raw. In recent years a restricted *V. parahaemolyticus* biovar (O4:K12, urease$^+$) has become dominant in coastal waters of the western US and Mexico and is also the major pathogenic strain. Dominance may reflect a more effective response to environmental pressures (Varnam and Evans, 1991). A further pathogenic species, *V. vulnificus*, which is also associated with shellfish, is amongst vibrios isolated from the stomach of finfish where the pH value is *c.*4.0. Numbers of up to 10^6 g^{-1} have been isolated from bottom-feeding fish consuming molluscs and crustacea. The presence of *V. vulnificus* in finfish is not considered to be of major public health significance, since such fish are cooked before consumption. Finfish may, however, be of importance in the introduction of *V. vulnificus* to new areas. Finfish are mobile, travelling in large schools and could have an immediate effect on water quality in shallow areas with little water movement (DePaola *et al.*, 1994).

Luminescent bacteria are also more widespread in neritic and littoral waters and the luminescent symbiosis more common. Luminescence is a relatively widespread trait in marine bacteria, which may be considerably more common than currently appreciated. The majority of marine luminescent strains are fermentative, Gram-negative rods of the

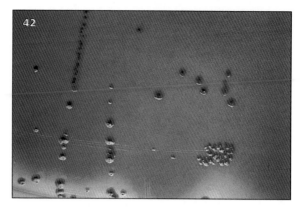

42 *Vibrio parahaemolyticus*. With the exception of *V. hollisae*, most vibrios grow well on TCBS cholera medium (originally designed for the recovery of *V. cholerae*). Sucrose-fermenting species, including *V. cholerae*, produce yellow colonies while non-sucrose-fermenting species, including *V. parahaemolyticus*, produce green colonies. (Source: seawater; Medium: TCBS cholera medium; Incubation: 37°C (98.6°F), 24 hours.)

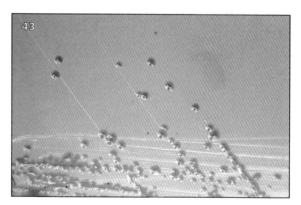

43 *Vibrio fischeri*. Like *V. parahaemolyticus*, *V. fischeri* does not ferment sucrose. The strain illustrated produces a yellow-brown pigment that obscures the green colour of the colonies. (Source: inshore seawater; Medium: TCBS cholera medium; Incubation: 37°C (8.6°F), 24 hours.)

genera *Vibrio* and *Photobacterium*, the sole exception being the non-fermentative *Alteromonas hanedai*. The best known luminescent bacteria are *V. fischeri* (**43**), *P. phosphoreum* and *P. leiognathi*, which form symbioses with fish and other marine animals. These bacteria are also free-living in the marine environment and present on the surface of fish as saprophytes. In addition no symbiotic relationships are known for common luminescent bacteria such as *V. harveyi*. In bioluminescent symbioses, the bacterial symbiont is harboured, in pure cultures, in specialized light organs and receives nutrients from the fish. The bacterium also benefits from exclusion of competitors and dissemination from the light organ into new

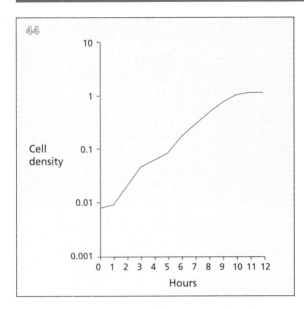

44 This illustration and the following two show the autoinduction of luminescence and luciferase synthesis in *Vibrio fischeri*. Although cell division recommences c. 1 hour after transfer of luminescent cells to a new culture, luciferase synthesis ceases for c. 4 hours.

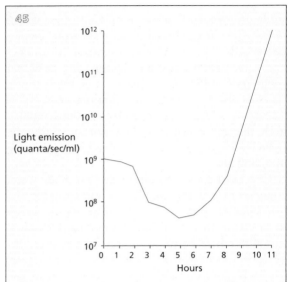

45 Cessation of luciferase synthesis leads to a marked decrease in luminescence during the first 4–5 hours after transfer.

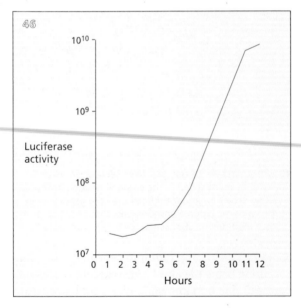

46 A high level of *de novo* synthesis of luciferase after 4–5 hours, when cell numbers are sufficient to permit accumulation of the autoinducer (**47**) to threshold levels. This leads to a dramatic increase in luminescence (**45**).

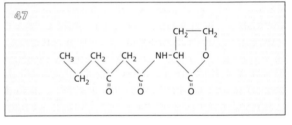

47 Structure of the luminescence autoinducer *N*-(3-oxo-hexanoyl) homoserine lactone.

habitats, such as seawater and the fish gut. Light produced by the bacterium is utilized by the fish in a number of ways, including feeding and, possibly, signalling and antipredation.

Available evidence strongly suggests that life in the light organ for the bacterium involves a low growth rate, 20 to 30 times slower than in laboratory culture, and high light production. In *V. fischeri* luciferase synthesis and the resulting luminescence are controlled by a regulatory mechanism known as autoinduction (**44–46**). Control is mediated by an autoinducer, *N*-(3-oxo-hexanoyl)homoserine lactone (**47**), which acts as a bacterial pheromone. Synthesis of luciferase and other luminescence-associated enzymes is triggered by the concentration of autoinducer reaching a threshold level, usually when cell numbers exceed 10^7 ml^{-1}. This means that the expression of luminescence in *V. fischeri* is frequently coupled with symbiosis since the autoinducer would not be expected to accumulate to threshold levels in seawater where cell densities are typically less than 10^2 ml^{-1}. The mechanism also ensures that the energy hungry process of bioluminescence is avoided, and energy conserved, when cell numbers are low.

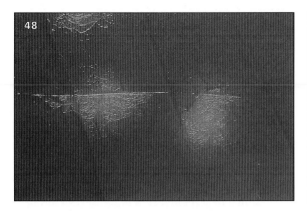

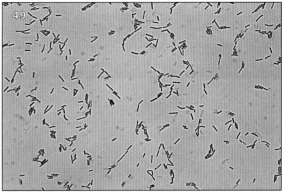

48 A halotolerant species of *Bacillus*. Although *Bacillus* is a ubiquitous organism, relatively little is known of isolates from marine sources. This isolate, which resembled the little-studied '*B. filicolonicus*', was capable of growth at NaCl concentrations in excess of 12%. (Source: seawater, English channel; Medium: blood agar; Incubation: 28°C (82.5°F), 24 hours. Note that the original isolation was made on seawater agar but colonial morphology is best seen on blood agar.)

49 Halotolerant *Bacillus* species. The positive Gram reaction has been virtually entirely lost. (Source: as for **45**; Microscopy: Gram stain, × 1,200.)

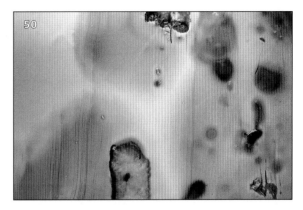

50 A gliding bacterium from marine mud. Addition of crystal violet to 3.5% NaCl medium to inhibit halotolerant *Bacillus* permitted development of a mixed microflora of Gram-negative, rod-shaped bacteria, including small, swarming colonies. Swarming was unaffected by the relatively high nutrient content of the medium or by the presence of crystal violet, but a minimum of 1% NaCl was required. (Source: intertidal mud flats; Medium: nutrient agar with 0.02% crystal violet and 3.5% NaCl; Incubation: c.18°C (64.4°F), 5 days.)

The environmental benefit of luminescence to free-living or saprophytic cells is less readily apparent. It may be a factor in enhancing dispersal since luminescent micro-organisms, growing saprophytically on dead fish, attain sufficient cell density to induce luminescence. This in turn is likely to attract marine organisms to the food source, resulting in propagation and dispersal (Meighen, 1992). However, non-luminescent bacteria appear to be equally successful in dispersal. A possible direct role for the luminescent system is the function of luciferase as an alternative electron carrier to the cytochrome system. This may enhance survival and growth at low oxygen tensions (Ulitzur *et al.*, 1981).

Some of the genera present in, and adapted to, coastal and estuarine waters are not present in oceans. These include *Bacillus* (**48, 49**), the micro-aerophil *Oceanospirillum*, appendaged bacteria including *Caulobacter*, unicellular gliding bacteria including *Cytophaga*, *Capnocytophaga* and strains of unknown identity (**50**). *Bdellovibrio* has been isolated from coastal seawater, but its significance is unknown. The gliding and highly differentiated bacterium *Leucothrix* is present as an epiphyte on seaweed, especially in intertidal zones. It is present in large numbers in decaying seaweed and important in its degradation. Marine strains of *Cytophaga* are also common and of importance in the degradation of organic matter, comprising up to 33% of the cellulolytic microflora. *Streptomyces* can also be readily isolated from coastal seawater, especially where nutrient concentrations are high.

The presence of this organism was, for many years, attributed to exospores being washed in from terrestrial sources but there is now evidence of an indigenous, marine origin (Moran *et al.* 1995).

Relatively little is known of micro-aerophilic bacteria in seawater, although both neritic and ocean waters would appear to offer favourable conditions for their growth. A study of the microflora of Tokyo Bay showed micro-aerophils to be present at a level of 50–100% of that of aerobes and facultative anaerobes, suggesting a role in mineralization. Nine phena were identified, resembling, but distinct from, non-marine genera (Sugita *et al.*, 1993). Magnetotactic, micro-aerophilic bacteria have been isolated from coastal environments (see pages 72–73).

Neritic and littoral sediments are recognized as sites of intense microbial activity. It was once thought that the bulk of degradation occurred in the sediment but there is now evidence that the importance of bacteria in the water column has been underestimated. A very high rate of deposition of organic matter means that depletion of O_2 within sediments occurs within a few millimetres of the surface (51). This may be contrasted with deep oceans where the rate of deposition is low and sediments are oxic for many metres below the surface. Bacteria are of the greatest importance in the sediment, anoxic mineralization often occurring at the same, or a higher, rate than the oxic process (52) (Hansen and Blackburn, 1992). The high quantity of sulphate ($c.28$ mM SO_4^{2-} vs. $c.0.2$ mM O_2) in sediments means that sulphate-reducing bacteria are of major importance and account for 70% (approx.) of mineralization in sediments. Some of the H_2S produced by sulphate reduction reacts with iron or manganese while the remainder diffuses upwards to the oxic zone. In the latter case, microbial mat communities may develop just within the sediment surface. These consist of vertically stratified communities and cover large areas of mudflats in salt marshes and the upper intertidal region of sandy beaches. The structure of the mat may vary (53, 54) but the dominant members are bacteria which can metabolize reduced inorganic sulphur compounds and elemental sulphur. Overlap occurs between H_2S and O_2 zones in sediments with

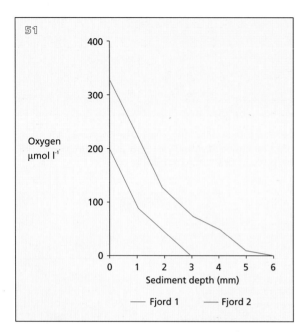

51 Development of anoxic conditions in fjord sediment at two depths. The rate of deposition of organic matter may be described as high throughout the fjord system. There is, however, a distinct difference between site 1, at which the depth of water was only 3.5 m, and site 2 at 220 m.

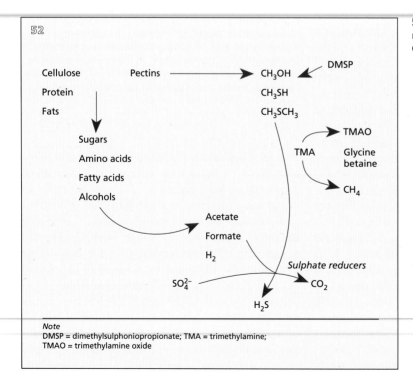

52 Anaerobic degradation of organic matter in coastal sediments. After Oremland (1988).

Note
DMSP = dimethylsulphoniopropionate; TMA = trimethylamine;
TMAO = trimethylamine oxide

high organic intensity and intense rates of sulphur reduction or in iron-depleted zones.

Aerobic chemolithotrophic autotrophs – primarily the highly H_2S-tolerant *Thiomicrospira* and possibly motile, filamentous types such as *Beggiatoa* which fix N_2 and CO_2 – proliferate at the H_2S–O_2 interface. Immediately beneath the surface of the H_2S zone is a layer of anoxygenic, purple sulphur bacteria, primarily *Chromatium* and *Thiocapsa* (55, 56), which anaerobically oxidize H_2S and fix CO_2 and N_2. Purple sulphur bacteria are able to coexist with colourless sulphur bacteria,

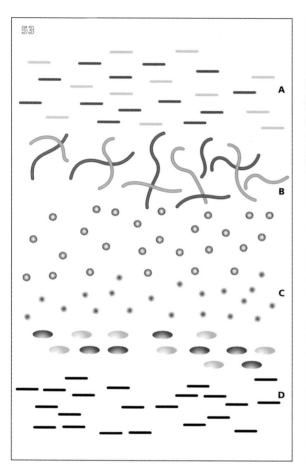

54 Vertical section through a beach mat. The upper layer consists of algae and cyanobacteria, below which is the pink-pigmented purple sulphur bacterium *Thiocapsa roseopersicina*. The black bottom layer is sulphide-containing sediment. Reproduced by courtesy of Prof. R.A. Herbert, University of Dundee.

53 Structure of the microbial mat community, Great Sippenwisset salt marsh, Massachusetts. *Structure of zones* A. Diatoms and cyanobacteria (*Navicula*, *Lyngbya*); photopigment chlorophyll *a* (680 nm). B. Cyanobacteria (*Microcoleus*, *Oscillatoria*); photopigment chlorophyll *a* (680 nm). C. Purple sulphur bacteria (*Thiocapsa roseopersicina*, *Thiocystis*); photopigment bacteriochlorophyll *a* (850 nm). Purple sulphur bacteria (*Thiocapsa pfennigii*); photopigment bacteriochlorophyll *b* (1020 nm). D. Sulphide-rich zone containing *Desulfovibrio*; non-photosynthetic. After Pfennig (1988).

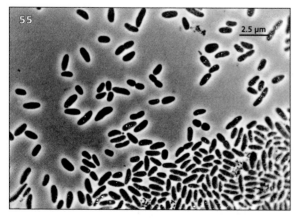

55 Purple sulphur bacteria *Chromatium salexigens*. Reproduced by courtesy of Prof. P. Caumette, University of Bordeaux.

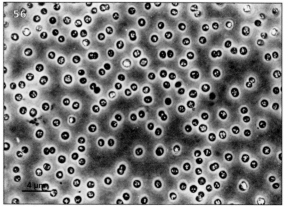

56 Purple sulphur bacteria *Thiocapsa halophila*. Reproduced by courtesy of Prof. P. Caumette, University of Bordeaux.

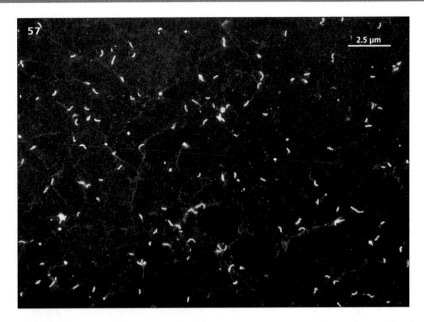

57 Dissimilatory sulphate reduction is one of the most important pathways in marine sediments for the final mineralisation of organic carbon. Despite this, the full metabolic potential of sulphate-reducing bacteria (SRB) has only been recently recognised. It is also apparent that the diversity of described species of SRB is rapidly increasing (Widdel and Bak, 1992). As in many other circumstances, viable counts may not reveal the true picture. Several distinct populations of SRB were detected in sediment from the German Baltic Sea using an indirect immunofluorescence technique. This is a photomicrograph of immunofluorescence exhibited by a pure culture of a vibroid SRB stained by the homologous anti-SRB antiserum raised for that isolate. Note the less strongly fluorescing flagella. Reproduced with permission from Lillebaek, 1995, © 1995 American Society for Microbiology.

despite the much greater affinity for sulphur of the latter. Purple sulphur bacteria, however, obtain electron donors from the products of incomplete sulphur oxidation by *Thiobacillus*. There may also be an underlying layer of green sulphur bacteria, which have similar biochemical activities. These include *Chlorobium* and *Prosthecochloris*. The bottom layer of the mat is a sulphide-containing sediment where variable numbers of sulphate-reducing bacteria, such as *Desulfovibrio* and *Desulphobacter* (**57**) are present. Surface mats usually have an upper layer consisting of cyano-bacteria (*Oscillatoria*, *Microcoleus*) possibly overlaid by eukaryotic phototrophs. Sulphide-tolerant cyano-bacteria may also develop alongside purple sulphur bacteria, the oxygen produced being involved in oxidation of H_2S. As noted earlier the reduction in the ozone layer and consequent increase in solar UV irradiation, especially UV-B, has a negative effect on cyanobacterial primary production, although the situation in mats is less well understood than with free-living micro-organisms. *Microcoleus chthono-plastes*, a very widely distributed mat builder, appears able to sense UV-B directly and to minimize exposure by migrating vertically downwards at periods of high UV intensity. By implication, the depths at which maximum photosynthesis and the maximum O_2 concentration occur are also displaced downwards. Despite this defensive strategy, there is a loss of overall productivity (Bebout and Garcia-Pichel, 1995).

Although reduction of sulphate has been thought to be restricted to the anoxic zone, radiotracer studies have shown that the process also takes place in the oxic zone of a cyanobacterial mat (Jorgensen, 1994). Sulphate-reducing bacteria were present which can respire, although not grow, with O_2 as electron acceptor and even oxidize reduced sulphur com-pounds. These bacteria are present throughout the oxic layer of mats. Simultaneous oxidation and reduction of thiosulphate occurred in all layers of the mat, although oxidation was predominant in the oxic layers and reduction predominant in the anoxic.

The location of phototrophs in distinct zones has been correlated with the spectral characteristics of the light penetrating the layers of the mat and the photopigments produced by the micro-organisms in each layer. Work with oxygenic, phototrophic organisms in a coastal sediment demonstrates that complementary light utilization is only one factor

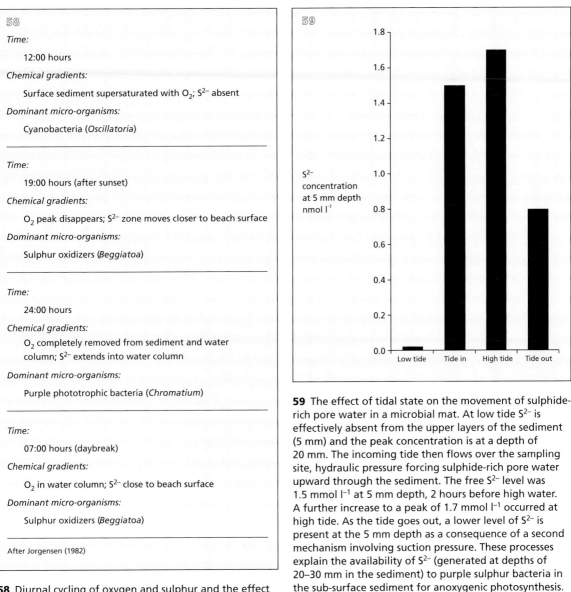

58

Time:

12:00 hours

Chemical gradients:

Surface sediment supersaturated with O_2; S^{2-} absent

Dominant micro-organisms:

Cyanobacteria (*Oscillatoria*)

Time:

19:00 hours (after sunset)

Chemical gradients:

O_2 peak disappears; S^{2-} zone moves closer to beach surface

Dominant micro-organisms:

Sulphur oxidizers (*Beggiatoa*)

Time:

24:00 hours

Chemical gradients:

O_2 completely removed from sediment and water column; S^{2-} extends into water column

Dominant micro-organisms:

Purple phototrophic bacteria (*Chromatium*)

Time:

07:00 hours (daybreak)

Chemical gradients:

O_2 in water column; S^{2-} close to beach surface

Dominant micro-organisms:

Sulphur oxidizers (*Beggiatoa*)

After Jorgensen (1982)

58 Diurnal cycling of oxygen and sulphur and the effect on the distribution of micro-organisms in a microbial mat community.

59 The effect of tidal state on the movement of sulphide-rich pore water in a microbial mat. At low tide S^{2-} is effectively absent from the upper layers of the sediment (5 mm) and the peak concentration is at a depth of 20 mm. The incoming tide then flows over the sampling site, hydraulic pressure forcing sulphide-rich pore water upward through the sediment. The free S^{2-} level was 1.5 mmol l^{-1} at 5 mm depth, 2 hours before high water. A further increase to a peak of 1.7 mmol l^{-1} occurred at high tide. As the tide goes out, a lower level of S^{2-} is present at the 5 mm depth as a consequence of a second mechanism involving suction pressure. These processes explain the availability of S^{2-} (generated at depths of 20–30 mm in the sediment) to purple sulphur bacteria in the sub-surface sediment for anoxygenic photosynthesis. From Herbert, 1992.

explaining zonation and that total light intensity and the chemical microenvironment are probably more important factors (Ploug *et al.*, 1993). Cyanobacteria, for example, are better adapted to low light intensities than overlying diatoms and are also of significantly greater H$_2$S tolerance.

Most surface sediments are highly dynamic habitats in which physicochemical conditions can undergo rapid change. The timescale can be very short and it has been demonstrated that the dissolved O_2 concentration in surface sediments dominated by cyanobacteria and algae changes measurably in response to an event as transitory as a passing cloud. In mat systems oxygen concentrations fall rapidly after sunset, leading to the movement of free S^{2-} to the sediment surface and then into the water column. This process, which is accompanied by the migration of micro-organisms and visible change of colour of the sediment and water, is reversed at dawn (58) (Jorgensen, 1982).

Factors other than light may be involved in determining physicochemical gradients. In a laminated beach system, the tidal state determined levels of S^{2-} in the surface layers of the sediment (59). This results in S^{2-}, generated at depths of 20–30 mm in the sediment, being made available to phototrophic purple sulphur bacteria resident at a much shallower depth of 1.5–2.0 mm. The cyclic nature of S^{2-} availability leads to the formation of blooms of *Thiocapsa* over a relatively short time scale (**60**, **61**) (van Gemerden *et al.*, 1989).

In sediments of lower organic input, the O_2 and H_2S zones do not overlap and are separated by a sub-oxic, or Fe–Mn zone. This sub-oxic zone is the site of anaerobic oxidation of organic compounds by Fe(III) and, less importantly, Mn(IV). H_2S reacts with Fe(II) to generate FeS which forms pyrites (FeS_2) by reaction with polysulphide. Pyrites is re-oxidized by O_2 on a seasonal basis. In coastal sediments, re-oxidation occurs in winter when O_2 penetrates deepest due to the reduced settlement of detritus. In contrast, re-oxidation in salt marshes occurs in summer when O_2 flows into the sediments, promoted by the higher level of photosynthetic activity. Pyrite accumulates in winter.

Methanogenesis is limited in marine environments by competition from sulphate-reducing bacteria. Methanogens, unlike sulphate-reducing bacteria, are able to metabolize trimethylamine derived from the osmolytes trimethylamine oxide and betaine. This pathway permits a certain level of methanogenesis, although in some cases it has been suggested that methanogenesis proceeds at a higher level than can be accounted for in this way. High levels of sulphate reduction occur in some situations including significant sewage pollution. Sulphate-reducing

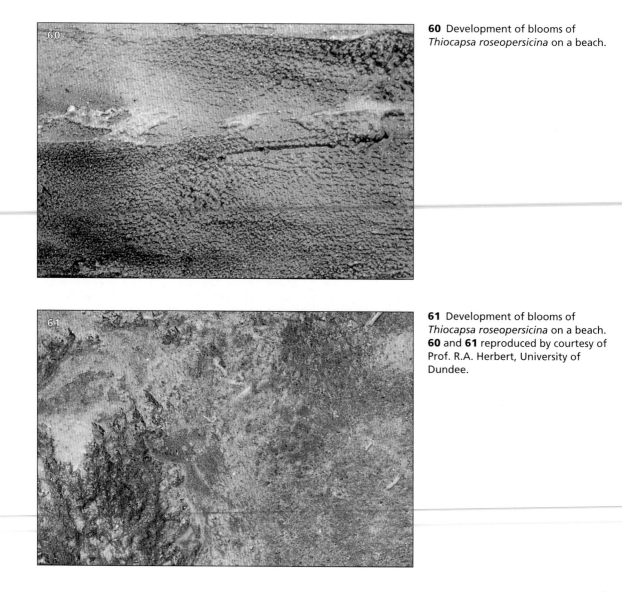

60 Development of blooms of *Thiocapsa roseopersicina* on a beach.

61 Development of blooms of *Thiocapsa roseopersicina* on a beach. **60** and **61** reproduced by courtesy of Prof. R.A. Herbert, University of Dundee.

bacteria develop in biofilms and a high level of activity leads to rapid corrosion of metal structures. The death of fish and marine animals may also result from incomplete oxidation of high levels of H_2S.

Bacteria in sediments use electron acceptors in a sequence dictated by energy yields. The sequence of oxidation is approximately reflected by the depths within the sediment at which different processes proceed at maximum rates (**62**). This sequence can, however, be disturbed by seepage of groundwater containing for example O_2 and NO_3 beneath the sulphate-reducing zone.

Although general patterns of the microbiology of neritic and littoral zones may be defined, it is important to appreciate that these areas support a wide range of habitats with specialized microfloras. On coral, for example, a horizontally migrating cyanobacterial mat develops. The mat is responsible for the economically important 'black band' disease of coral (**63**) (Carlton and Richardson, 1995).

62 Distribution of electron acceptors for bacterial metabolism in sediments.

62		
	Water	
1. Organic C + O_2	$\rightarrow$	$CO_2 + H_2O$
2. Organic C + NO_3	$\rightarrow$	$CO_2 + N_2$
3. Organic C + NO_3	$\rightarrow$	$CO_2 + NH_4^+$
4. Organic C + MN(IV)	$\rightarrow$	CO_2 + MN(II)
5. Organic C + Fe(III)	$\rightarrow$	CO_2 + Fe(II)
6. Organic C + SO_4^-	$\rightarrow$	$CO_2 + S$
7. Organic C	$\rightarrow$	$CO_2 + CH_4$

63 Coral is of geological importance in favouring the development of atolls and protecting coastlines. It is a habitat for a range of life, including reef dwelling fish. Some types of coral are used to make decorative items and can be of economic significance to local communities. 'Black band' disease is aptly named, the band being seen clearly between the lower living coral and the upper dead material (the nail provides a perception of scale). The cause appears related to sulphide production by a cyanobacterial mat. This resembles other cyanobacterial mats in structure but is unusual in migrating horizontally rather than vertically. Reproduced with permission from Carlton and Richardson, 1995, © Federation of European Microbiological Societies.

Consequences of salinity changes

A feature of neritic and littoral habitats is that the salinity is subject to marked and, in some cases, sudden variation. For this reason environments of this type offer a considerable challenge both to marine and freshwater micro-organisms although the two 'types' are able to coexist. The particular circumstances affecting micro-organisms are the consequence of the proximity of the terrestrial environment and fresh water entering the sea through rivers, drains, and flooding with consequent localized dilution. Estuarine environments are particularly subject to changes in NaCl concentration through changes in river flows, tidal patterns, and the like. At the same time, evaporation of water in hot weather means that NaCl concentrations in localized parts of intertidal habitats can rise significantly under some tidal conditions.

Protection against osmotic stress is obtained by the accumulation of a restricted range of low molecular weight organic compounds (compatible substrates). *Thiocapsa* from marine mats, for example, accumulates K^+ and sucrose. Glycine betaine is important in halotolerant and halophilic bacteria but is not accumulated by other bacteria. There is also evidence that slime offers protection, probably through maintaining a microenvironment around the cell and preventing sudden osmotic change (**64, 65**).

Most of the bacteria and cyanobacteria present in environments of fluctuating salinity are able to grow over a wide NaCl range (**66**). Some may originally be of terrestrial origin (see above) while many marine bacteria have only a very low requirement for Na^+ or other ions. There may, however, be physiological or morphological changes (**67, 68, 69**) as a consequence of growth at different

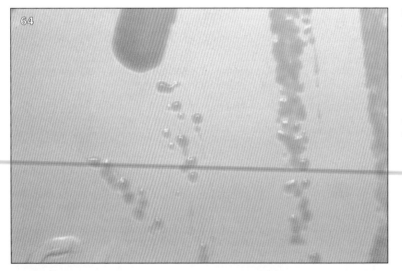

64 Effect of osmotic shock on a marine *Vibrio* species. The organism, isolated from seashore mud, produces copious slime when grown in liquid culture, adhering strongly to the walls of the culture vessel and increasing the viscosity of the medium. This is reflected in the appearance of colonies on seawater agar.

65 Cells (1) transferred from a medium containing 4% NaCl to water retained both integrity and viability. In contrast, cells (2) from a mutant, non-slimy strain of the same *Vibrio* lysed immediately on transfer to water.

66 An unidentified, halophilic, Gram-negative, rod-shaped bacterium. The bacterium, which differed from known genera, was able to grow at NaCl concentrations ranging from 0.5 to more than 25% added NaCl. The optimum is 7.5–10%. (Source: intertidal muddy pool; Medium: tryptone–soya agar + 7.5% NaCl; Incubation: c.18°C (64.4°F), 7 days.)

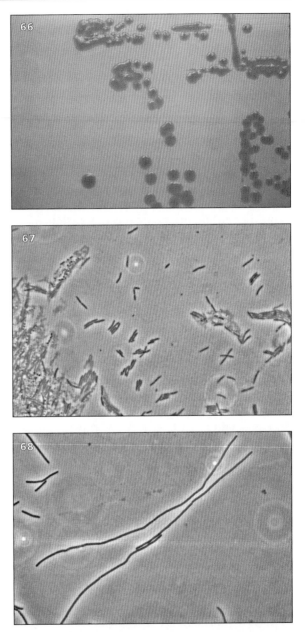

67 Effect of NaCl concentration on cellular morphology of an unidentified, halophilic, Gram-negative, rod-shaped bacterium. At NaCl concentrations of 7.5–25% the cells are fairly regular although the arrangements of some groups superficially resemble those of 'coryneform' bacteria. (Source: as for **66**; Microscopy: phase contrast, × 1,100.)

68 Effect of NaCl concentration on cellular morphology of an unidentified, halophilic, Gram-negative, rod-shaped bacterium. At lower NaCl concentrations than in **67** (5.0%), a few isolated cells of extreme length are present. (Source: as for **66**; Microscopy: phase contrast, × 1,100.)

69 Effect of NaCl concentration on cellular morphology of an unidentified, halophilic, Gram-negative, rod-shaped bacterium. At 2.5% NaCl, the majority of the cells are of extreme length. Despite the elongation there are no other indications of ionic stress affecting the cells. (Source: as for **66**; Microscopy: phase contrast, × 1,100.)

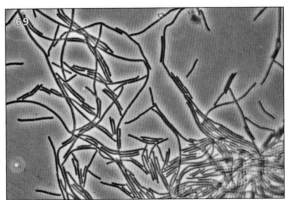

Na⁺ concentrations. In many situations bacterial numbers are greatest at intermediate salinities. These may represent a point in a succession between fresh and marine water environments where both freshwater and marine bacteria can adapt to growth. It has also been suggested that the high numbers may partly result from an alleviation of grazing pressure by protozoa. This is a consequence of the fact that hypertonic freshwater and hypotonic marine protozoa are unable to survive even relatively small increases or decreases in salinity (Bordalo, 1993). Zones of intermediate salinities may also be important in the establishment of epidemic foci by the pathogen *Vibrio cholerae*.

Consequence of sewage pollution

Coastal and estuarine waters may receive significant quantities of both raw and wholly, or partially, treated sewage. In some areas considerable quantities of sewage sludge are disposed of by dumping in deeper waters. There has been considerable concern over sewage pollution of beaches. In addition, pollution of water used for mariculture introduces the risk of enteric pathogens re-entering the human food supply. At the same time sewage is recognized as being responsible for increases in nutrient loading and thus associated with eutrophication.

Sewage is a potential source of a very wide range of human pathogens, although in practice illness is caused by a more restricted range. Members of the *Enterobacteriaceae* such as *Salmonella* and the index organism *Escherichia coli* have been of particular concern and many studies have been made to determine survival in seawater (*Feature 5*). In the past, the survival of these organisms has been considered to be short, the major factors leading to death being visible and UV light, salinity, starvation, low temperature and predation. These in turn are influenced by the physiological condition of the cell, sensitivity decreasing with its age from lag through stationary phases and being greater with aerobically than anaerobically grown cells (Gauthier *et al.*, 1992). Many studies have been based on loss of culturability and little account has been taken of the viable nonculturable state. In artificial seawater, *E. coli* was shown to maintain culturability for at least three years and to selectively adapt to salinity (Byrd and Colwell, 1993). *E. coli* also retains plasmids under these conditions; this has implications for the release of genetically modified micro-organisms.

Feature 5. Come on in, the sea's lovely

Despite the denials of some tourist authorities, sewage pollution of near-shore environments is a risk to public health in a number of ways. Enteric bacteria have been associated with food poisoning following the consumption of raw shellfish grown in polluted waters. There also now appears to be strong evidence of a higher incidence of enteric illness amongst bathers on heavily polluted beaches. Viruses are discharged with sewage, and infections with enteroviruses, including poliovirus, have resulted from bathing in contaminated seawater. In recent years, the increased consumption of raw shellfish has been a factor in a number of large-scale outbreaks due to small, round, structured viruses (SRSVs) and smaller outbreaks of hepatitis A and E infections.

In warmer climates at least, wound infections are both more common and more likely to result in serious illness or death (Kueh *et al.*, 1992). These can involve indigenous bacteria, such as *V. vulnificus*, but members of the *Enterobacteriaceae* are also associated with localized wound infections. *Aeromonas hydrophila* is also frequently associated with wound infections following seawater exposure, some of which may be severe. In many countries, the surfing community has played a major role in campaigning for less polluted beaches and their adjacent waters, forming associations against pollution.

Sea ice

On a seasonal basis, *c.*12% of the Earth's surface is layered by sea ice where sea ice microbial communities (SIMCOs) develop, usually in microscopic brine pockets. These are of variable salinity (up to five times that of surrounding seawater), have a temperature of −2 to 0°C (28.5 to 32°F) and a light intensity of 0.02–0.80% of that at the surface (Gosinski *et al.*, 1993). Communities are dominated by algae which contribute significantly to primary production in polar habitats, but protozoa and bacteria are also present. Extremely psychrophilic, pigmented, gas-vacuolated bacteria of unknown identity have been isolated from both arctic and antarctic sea ice. Freshwater snow and ice covering high-pH mountain lakes also support highly active microbial communities (see pages 70–71).

THE FRESHWATER ENVIRONMENT

Introduction

Despite some general similarities with the marine environment, freshwater habitats are generally more favourable and there is certainly no parallel with the energy-poor, cold and barophilic conditions present in the ocean deeps. Nutrient levels in freshwater habitats vary but are usually significantly higher than in oceans. Bodies of fresh water receive higher levels of organic matter both from runoff from land and through closer contact with the works of man. The relatively small volume of freshwater bodies also means that pollution with toxic substances is potentially a greater problem. Thermal pollution, due to the discharge of cooling waters and other industrial processes, can also have a significant effect on the local microflora.

Running water

Running water includes a number of different situations. The term is used equally to describe a fast-flowing mountain stream, a sluggish, heavily polluted ditch, or a great lowland river. The microbiology of apparently very different types of running water is often surprisingly similar in a general sense. Superimposed on the similar general pattern are differences often involving the presence or absence of specific groups of micro-organisms.

In the case of streams and rivers, it is common to imagine a continuum from the source to, ultimately, the sea in which the initially pristine water acquires an increasing load of nutrients, pollution and micro-organisms. To some extent this is true (70, 71) but great care must be taken to avoid over-

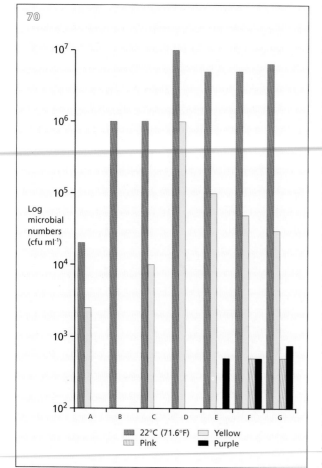

70 Development of the microflora of the River Kennet in southern England from close to its source to the confluence with the River Thames. The River Kennet arises as two separate branches in chalk uplands. During the summer months, the channels at the source are often dry and flow is limited in the upper reaches. The river flows in a generally eastward direction to the River Thames, the total length being approximately 103 km (64 miles). The Kennet and Avon canal enters the Kennet some 32 km (20 miles) from the source and thereafter the river is canalized (the 'Kennet navigation') and consists of both natural river and stretches of navigation cuts. There are two significant tributaries, the River Lambourne and the River Enbourne, both of which enter in the eastern half. The river system is further complicated by a large number of side channels, fish ponds and disused, but watered, mill leats. Some of the side channels are of very high microbial productivity. The river is subject to only limited chemical or thermal pollution although it receives treated sewage effluent at a number of points. The largest volume input is close to the confluence with the River Thames, but in terms of water quality, the most significant is probably the first discharge some 10 km (6 miles) from the source. There is also evidence of pollution from agriculture and fish farms. The general water quality is, however, considered to be good. The river supports a variety of fish and otters have recently returned.

Over a period of two years, water was sampled at a number of points between the first, close to the source and the confluence with the River Thames. Basic microbiological parameters were determined: a 'total viable count' on quarter-strength nutrient agar and differential counts based on colony pigmentation, i.e.

simplification. Microbial numbers may increase fairly rapidly even when the source is a spring of very high purity. In some cases this has been attributed to rapid growth of autochthonous micro-organisms at higher temperatures and in relatively nutrient-rich conditions. Runoff from ground surrounding the watercourse is probably a more important factor, contributing both micro-organ-isms and nutrients. The contribution is likely to be variable depending on the nature of the land and any agricultural activities. Farm animals are of particular importance as a potential source of faecal bacteria and zoonotic pathogens (Fernandez-Alvarez *et al.*, 1991). It has also been suggested that wild animals and birds may be an important source of pathogens, including *Campylobacter*. Bacteria

(**70** *continued*)

non-pigmented, yellow, pink or purple. Gram stains showed each pigmented category to consist of Gram-negative, rod-shaped bacteria. The majority of non-pigmented bacteria were also Gram-negative rods although, on some occasions, Gram-positive rods with a 'coryneform' morphology were also present in significant numbers. While examinations of this kind may be regarded as very simplistic there is a lesson here that is not too late to learn: providing that the limitations of the methodology are recognized (and this is particularly so in the case of 'total viable counts') simple observations can yield a great deal of information on which more sophisticated work can be based.

The illustration shows the distribution of the microflora during the summer and autumn months (May to September). Yellow-pigmented, Gram-negative rods are the only pigmented bacteria present at the first sampling point near the source (A). Total numbers are much higher at the second point (B), which is just downstream from the first sewage discharge, but yellow-pigmented bacteria were never isolated at this site. The reason for this is obscure. Yellow-pigmented rods were again present in large numbers at the third sampling point (C) although 'total' numbers showed no increase. Both 'total' numbers and yellow-pigmented bacteria reach a numerical peak at the fourth sampling point (D). This may be attributed to water entering from a fish farm and from side channels of very high microbial productivity. The water temperature at this point was also about 1.5°C (2.7°F) higher throughout the year. The reason for this is not known. 'Total' numbers are somewhat lower at the next two sampling points, although an increase occurs at the final point (G). A marked fall in numbers of yellow-pigmented bacteria occurs between sampling points 4 and 5 (E) and continues to the final point. Purple-pigmented bacteria appear in small numbers at the fourth sampling point and pink-pigmented at the fifth (F). The overall pattern suggests that yellow-pigmented bacteria are favoured by low nutrient levels, decreasing in numbers as the organic content of the water rises. A marked decrease is apparent at the fifth sampling point, which may be at least partly attributed to input from extensive reed beds. The higher nutrient level does appear to favour purple- and pink-pigmented bacteria although total numbers of these remain small.

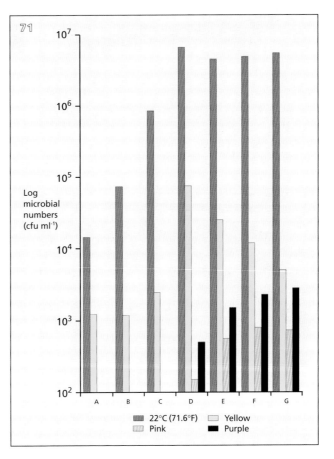

71 The development of the microflora in the winter and spring months shows a similar overall pattern although there are significant differences. 'Total' numbers of bacteria are rather lower, probably as a consequence of water temperatures. The effect on yellow-pigmented bacteria is particularly marked, numbers being significantly lower at all sampling points, except, paradoxically, the second sampling point where isolation is possible in winter but not summer. Purple- and pink-pigmented bacteria were also more prevalent in winter than in summer. First recoveries were made significantly higher up the river at the fourth sampling point and numbers, particularly of purple-pigmented, markedly higher. This may reflect a relatively greater ability to grow at the lower temperatures, although these bacteria may also have been favoured by the higher organic content of the river resulting from increased runoff and sediment entering the water as a result of turbulence.

able to multiply in moving water develop preferentially as biofilms although there is also colonization of particulate organic matter and aquatic plants. Many of the bacteria present in the water current have been shed from biofilms and are able to rapidly colonize new surfaces (72, 73). A wide variety of bacteria are present in freshwater including species of *Pseudomonas* (74–79), *Flavo-*

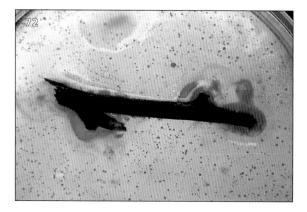

72 Colonization of a twig, freshly fallen into a small stream, commenced at points where flow was less strong.

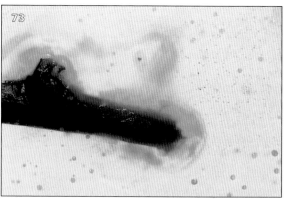

73 Colonizing bacteria were those most common in the general aquatic environment.

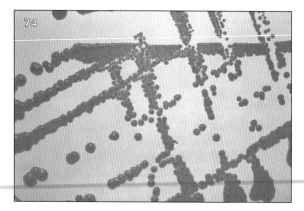

74 *Pseudomonas fluorescens*. Many strains of *Ps. fluorescens* isolated from water, soil and foods in Europe are non-pigmented. Most US isolates, however, produce fluorescent pigments. (Source: mountain stream; Medium: quarter-strength nutrient agar; Incubation: 22°C (71.6°F), 5 days.)

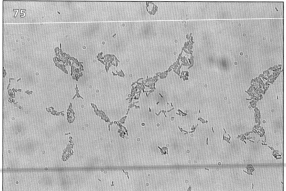

75 *Pseudomonas fluorescens*. A typical Gram-negative, rod-shaped bacterium; the cells of this strain are larger than most. (Source: as for **74**; Microscopy: Gram stain, × 1,200.)

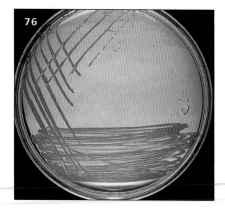

76 *Pseudomonas aeruginosa* is common in natural waters although usually present in small numbers. The ability of this organism to grow at relatively high temperatures, and its resistance to antimicrobial agents, enables it to attain high numbers (up to 10^8 cfu ml^{-1}) in jacuzzis and whirlpool baths. Pigment production is slow and not visible in the young culture under the lighting conditions used. Pigment production increases as senescence approaches and colonial appearance becomes highly characteristic. Appearance is usually related to strain. (Source: lowland river; Medium: nutrient agar: Incubation: 35°C (95°F), 24 hours.)

bacterium (**80**, **81**), *Alcaligenes*, *Cytophaga* and 'coryneforms'. These bacteria are consistently present in significant numbers in running water of all types. Many are pigmented, including members of genera normally considered non-pigmented. As in the marine environment, pigmentation is considered to impart protection against UV irradiation. The presence of dissolved organic compounds in freshwater also minimizes the effects of UV irradiation. Members of the *Enterobacteriaceae* can also be isolated from running waters in significant numbers. These bacteria have been associated with

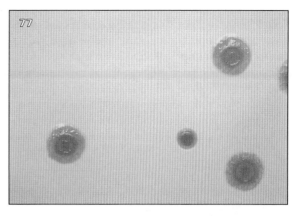

77 *Pseudomonas aeruginosa*. (Source: lowland river; Medium: nutrient agar: Incubation: 35°C (95°F), 24 hours.)

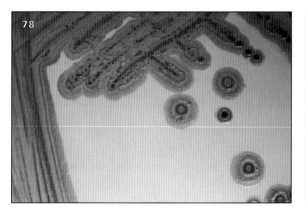

78 *Pseudomonas aeruginosa*. (Source: lowland river; Medium: nutrient agar: Incubation: 35°C (95°F), 4 days.)

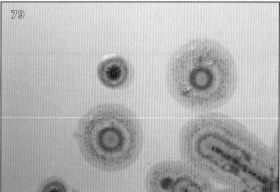

79 *Pseudomonas aeruginosa*. (Source: lowland river; Medium: nutrient agar: Incubation: 35°C (95°F), 4 days.)

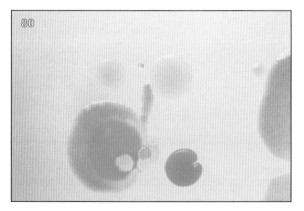

80 A putative *Flavobacterium*. Although many aquatic isolates are yellow pigmented, this isolate was unusual in that the slime was pigmented but not the colony itself (cf. **81**). Visible slime was not produced until the colony was fully developed. (Source: water, emerging through slime layer at outlet of clay drainpipe; Medium: quarter-strength nutrient agar; Incubation: 22°C (71.6°F), 6 days.)

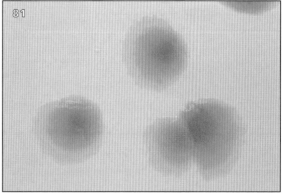

81 A putative *Flavobacterium*. The isolate was phenotypically similar to, but distinct from, **80**. The colony itself is pigmented but slime and spreading growth at the periphery is, initially, colourless. (Source: ditch receiving water from land drain; Medium: quarter-strength nutrient agar; Incubation: 22°C (71.6°F), 6 days.)

faecal pollution but this is a misconception in the case of environmental strains including *Enterobacter*, *Hafnia* and *Serratia* (82) which should be considered as a part of the normal microflora of freshwater. *Proteus* (83, 84) is also common and may be present in high numbers in water of high organic content. Some members of the *Enterobacteriaceae*, such as *Yersinia intermedia* (85), may originally be of animal origin but are able to adapt to aquatic life. Some members of the normal microflora are present only in small numbers and may require special techniques for isolation. These include the Actinomycetes, *Actinoplanes* (86) and *Micromonospora* (87–89) and the gas vacuolated bacterium *Ancyclobacter aquaticus* (*Microcyclus aquaticus*) (90).

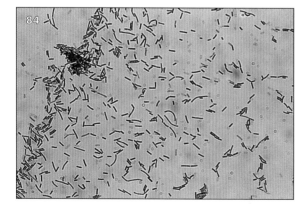

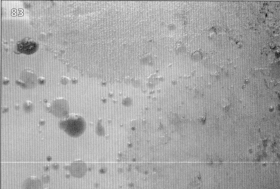

82 A prodigiosin-producing strain of *Serratia marcescens*. Pigmented strains of *S. marcescens* may synthesize either the non-diffusible red pigment prodigiosin or the diffusible pink pigment pyrimine. Non-pigmented strains are also common and form the majority of isolates from clinical material. (Source: chalk stream, lightly polluted by agricultural activities; Medium: nutrient agar; Incubation: 30°C (86°F), 2 days.)

83 A species of *Proteus*. The strain swarms on solid media but the periodicity normally associated with swarming in this organism is absent. (Source: silted side channel, lowland river; Medium: nutrient agar; Incubation: 30°C (86°F) , 2 days.)

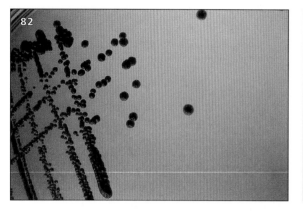

84 A species of *Proteus*. (Source: as for **83**; Microscopy: Gram stain, × 1,200.)

85 *Yersinia intermedia*. 'Environmental' strains of *Yersinia* are not uncommon in fresh water, although enrichment is usually required for their isolation. The pathogen *Y. enterocolitica* has also been isolated from water although its presence is thought to be due to contamination from animal sources, especially pigs and rats. (Source: lowland, moderately polluted river; Medium: cephaloridin–irgasan–novobiocin agar; Incubation: 28°C (82.4°F), 24 hours.)

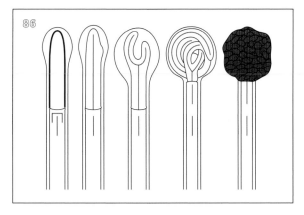

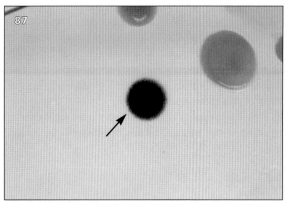

86 The sporangium of *Actinoplanes* arises by growth and coiling of the sporangial mycelium within the sporangium wall. The process is thus analogous to that of the streptomycetes, the sporangial wall serving the same function as the fibrous sheath of the streptomycete aerial mycelium. Aerial mycelia appear to be advantageous only in terrestrial habitats when dispersal of spores is by air. The mechanism of spore-bearing in *Actinoplanes* and *Micromonospora* is presumably better adapted to dispersal in water.

87 A species of *Micromonospora*. The organism (arrowed) is present in large but apparently variable numbers in river systems. Recovery can be enhanced by pretreatment of the sample at 80°C (176°F). Colonies can be mistaken for moulds in young cultures. (Source: river water, stored for four weeks at ambient temperature; Medium: nutrient agar: Incubation: 25°C (77°F), 5 days.)

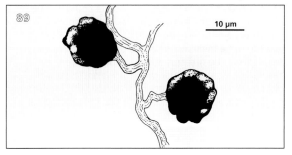

89 *Actinoplanes* and other actinoplanetes form from one to more than 1,000 spores. These are usually motile, within a sporangium arising directly from the substrate mycelium.

88 Although rRNA sequencing indicates that *Micromonospora* is closely related to other actinoplanetes, the developmental pattern is different. *Micromonospora* forms single spores, which are not enclosed in a sporangial wall, on the tips of hyphae within the substrate mycelium (× 2,300).

90 *Ancyclobacter aquaticus* (*Microcyclus aquaticus*). The translucent appearance of the colonies is indicative of a low incidence of gas vacuoles. Where gas vacuoles are abundant, the colonies have a chalky white, opaque appearance. (Source: river water; Enrichment: 100 ml water supplemented with 10 mg peptone, incubated at room temperature for two to three weeks; Medium: half-strength nutrient agar + 0.1% peptone; Incubation: 30°C (86°F) , 7 days.)

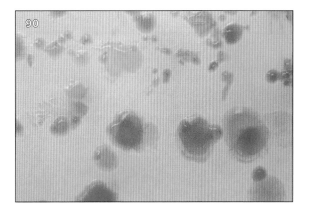

Species of *Aeromonas* (**91**) are present in highly variable numbers that may correspond closely to the trophic state of the water, while *Chromobacterium fluviatile* (**92**) appears to have a highly variable distribution, even over a small geographical area of the same water system.

A wide range of other bacteria may be found in running water but are more common in ponds and lakes. These include *Cytophaga* and other gliding bacteria, *Serpens* and 'non-motile, Gram-negative, curved bacteria' including *Flectobacillus*. A bacterium of the same type, *Spirosoma*, was, however, isolated from a canalized river in close proximity to a similar bacterium that had colonized lock gates at the water level (**93–95**).

A number of prosthecate bacteria can be isolated

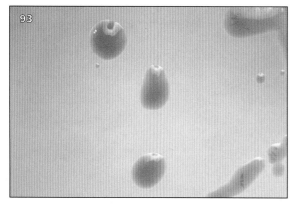

91 A species of *Aeromonas*. The isolate could not be identified with any existing species of *Aeromonas*. This is not an uncommon phenomenon with environmental genera, some members of which are pathogens, and results from the greatest attention being given, quite naturally, to those isolates which are associated with disease. (Source: side channel, canal; Medium: *Aeromonas* medium (Oxoid); Incubation: 28°C (82.4°F), 24 hours.)

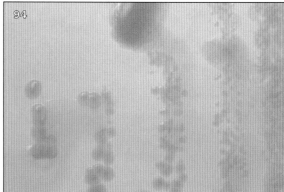

92 *Chromobacterium fluviatile*. Isolation could be made consistently from one area of an inland river navigation but not from other areas. Other colonies on the primary isolation plate were common waterborne bacteria. (Source: canal; Medium: quarter-strength nutrient agar; Incubation: 22°C (71.6°F), 3 days.)

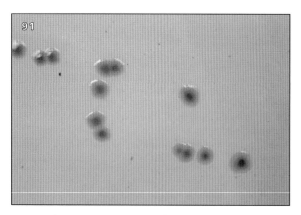

93 *Spirosoma linguale* isolated from freshwater by repeated streaking on MS medium. Presumptive identification may be made on the basis of phase contrast microscopy. (Source: bypass weir, canal; Medium: MS medium; Incubation: room temperature, 10 days.)

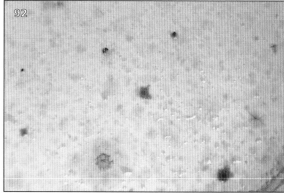

94 A slime-producing, *Spirosoma*-like organism. Isolation was made from a layer of slime just above the water level of lock gates. Despite the different colonial appearance to typical strains of *Spirosoma*, the cellular morphology, biochemical features and enzyme profiles (API-zym®) were very similar. *Spirosoma*-like organisms have also been isolated from gelatinous deposits on wet planks in a mine. The slime may have a role in protection from dehydration. (Source: slime, lock gates; Medium: MS medium; Incubation: room temperature, 10 days.)

from freshwater. These bacteria are usually present in truly oligotrophic environments, where nutrient levels are very low. In genera, such as *Hyphomicrobium*, the prosthecae serve a reproductive function, but in others the significance is less clear. In bacteria with long appendages, such as *Ancalomicrobium* (**96**) it may be postulated that the prosthecae impart a selective advantage by increasing the efficiency of nutrient uptake by increasing the surface area of the cell and retarding sedimentation.

In some cases, such as *Caulobacter*, the length of the prosthecae increases in response to low nutrient levels. A further feature of some prosthecate bacteria is the possession of a holdfast, usually situated at the tip of the prostheca, or the tip of the cell body. It is thought probable that bacteria with holdfasts, such as *Caulobacter*, preferentially attach to algae, protozoa, cyanobacteria and larger bacteria (**97**) and obtain nutritional benefit from organic secretions from the larger cell. Attachment to inanimate surfaces may also be advantageous nutritionally.

Sheathed bacteria are also a feature of freshwater environments and are often attached to solid substrates by a holdfast. *Leptothrix* is usually found

95 Lock gates on a canal. A thick slime develops at the waterline on the gates and other wooden canal structures. A mixture of Gram-negative bacteria is present, dominated by the *Spirosoma*-like organism.

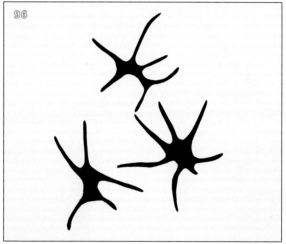

96 Cells of the prosthecate bacterium *Ancalomicrobium* (× 2,500).

97 *Caulobacter* attached to cells of a larger bacterium (× 4,000).

in unpolluted waters containing metal salts and accumulates ferric and manganic oxides in the sheath. *Sphaerotilus* (98) also accumulates metal oxides but is found in slow-running streams and ditches heavily contaminated with sewage or other organic matter.

Gallionella (99–101) is another structurally distinctive bacterium which, like *Sphaerotilus* may be referred to as an 'iron bacterium'. These bacteria, together with *Leptothrix* and *Sphaerotilus*, promote the corrosion of cast iron and mild steel pipes as well as of austenic stainless steel by the formation of ferric

98 *Sphaerotilus*, showing the sheath enclosing a chain of cells.

99 Unstained smear of *Gallionella* taken from iron- and manganese-bearing water from a stream flowing from an abandoned mine shaft. Photographed using light microscopy.

100 The same growth of *Gallionella* as in **99** but photographed using Nomarski interference microscopy. The spiral stalk of *Gallionella* is readily visible (arrowed).

101 When grown in liquid medium containing ferrous sulphide, *Gallionella* forms 'cottony' colonies attached to the sides of the vessel. These are very distinctive and an effective means of presumptive recognition of the bacterium. **99–101** reproduced with permission from Verran *et al.*, 1995, © 1995 Society for Applied Bacteriology.

hydroxide tubercles (Stott, 1988). *Gallionella* can also cause severe fouling of water wells but, as has been noted earlier, nothing is simple in the microbial world and the organism may also be used to remove ferrous iron from shallow aquifer supplies, thus permitting use for domestic purposes (Malkii, 1988).

Primary production in freshwater systems is photosynthesis by higher plants, eukaryotic algae and cyanobacteria, with higher plants being of much greater overall significance than in seawater. The majority of freshwater bacteria are heterotrophic and involved in the degradation of organic matter.

Moulds, especially the aquatic *Phycomycetes* (water moulds) (**102, 103**), are early colonizers of plant and animal materials and are involved in the degradation of complex molecules. In northern temperate waters, leaf material is an important energy source and the hyphomycete *Anguillospora longissima* is of considerable importance in the incorporation of leaf material into the detrital food web (Bermingham *et al.*, 1995). In many running waters the rate of organic turnover is high and mineralization takes place in the water column. Bacterial populations are primarily limited by protozoan predation and

102 Aquatic phycomycetes. (Source: woodland stream containing a high level of fallen twigs and branches; Medium: nutrient agar; Incubation: 20 °C (68°F), 5 days.)

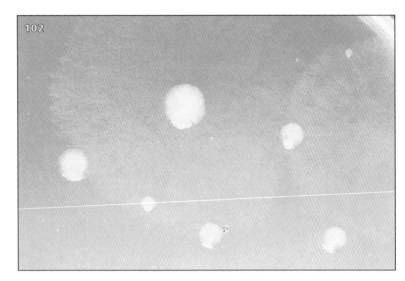

103 Life cycle of chytrids. Chytrids, which include the aquatic phycomycetes, produce motile spores and are the fungi most closely resembling the protozoa. The flagellated zoospore settles on a solid surface and development commences with the production of a branching system of rhizoids. Subsequent growth produces a spherical zoosporangium that cleaves internally to produce zoospores. These are finally liberated by the rupture of the zoosporangium to repeat the cycle.

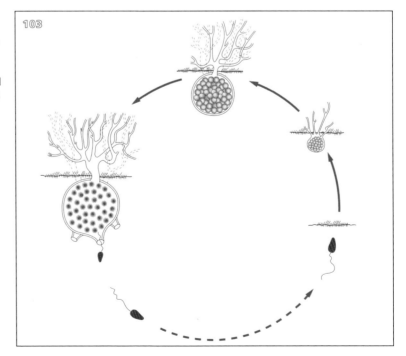

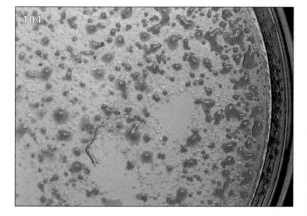

104 Confluent plaques on an isolation plate. The plaques were the result of cell lysis by presumptive *Bdellovibrio bacteriovorus*. Protozoa and myxobacteria were eliminated by phase contrast microscopy and the delayed development of plaques, together with increasing size during continuing incubation, eliminated bacteriophage. (Source: lowland river, about 250 m downstream of sewage works discharge; Medium: quarter-strength nutrient agar: Incubation: 25°C (77°F), 7 days.)

bacteriophage activity but the parasitic *Bdellovibrio bacteriovorus* can also be present (**104–106**). The relative importance of *Bd. bacteriovorus* in predation is a matter of debate but in overall terms it has been considered to be very limited. *Bd. bacteriovorus*, however, requires a high prey concentration ($c.10^6$ cells ml^{-1}) for survival and it is likely to be concentrated in small ecosystems where its importance may be very much greater. Although gliding myxobacteria are primarily soilborne, the nonfruiting algicidal genus *Lysobacter* (**107**) is present in slowly running water containing large quantities of plant debris (**108**). In some circumstances *Lysobacter* may limit the extent of algal and cyanobacterial blooms.

The extent of deposition of organic matter into sediment varies according to the nutrient load of the water system. In some cases, turbulent flow produces sufficient scouring to maintain most of the sediment in suspension. Particles are subject to bacterial degradation but may be sufficiently large

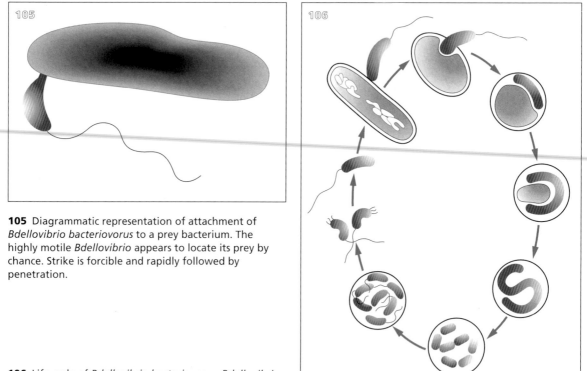

105 Diagrammatic representation of attachment of *Bdellovibrio bacteriovorus* to a prey bacterium. The highly motile *Bdellovibrio* appears to locate its prey by chance. Strike is forcible and rapidly followed by penetration.

106 Life cycle of *Bdellovibrio bacteriovorus*. *Bdellovibrio* adheres to the prey cell wall by its non-flagellate pole. As it penetrates the cell wall and settles in the periplasmic space, the prey cell becomes rounded, resembling a spheroplast. The *Bdellovibrio* cell continues to elongate until the nutrients derived from the shrinking prey protoplast are exhausted. *Bdellovibrio* then undergoes multiple division to uniformly sized cells, followed by lysis of the prey cell wall and release of free-living progeny. The parasitic life of *Bdellovibrio* is considered to be an extreme example of adaptation to environments of low nutrient concentration and offers a striking contrast with the more generally used strategies.

Feature 6. A lost wilderness

It is well known that the juxtaposition of large human populations with sensitive ecosystems leads to a high level of anthropogenic stress. The Everglades of Florida provide an example of anthropogenic stress affecting life of all types. Residents of, and visitors to, the Everglades must be aware of the commercial development of this wetland ecosystem, especially the all too visible out-of-town shopping malls. The micro-environment is adversely affected by nutrient enrichment, especially of phosphates, from intensive agriculture, altered hydrology as a result of flood control and water abstraction and relatively high mercury levels in the biomass. Sediments emit significant quantities of the 'greenhouse' gases methane and nitrous oxide. This may be a consequence of the nutrient level. In the Everglades, sediments close to agricultural runoff are characterized by low redox potentials, high phosphate concentrations and high water column conductivity. The number of anaerobic micro-organisms is 10^3 to 10^4 times greater than that from sites remote from agricultural runoff. This is probably a consequence of the high speed of response by anaerobes to an influx of nutrients such as phosphates, sulphates and nitrates. Data from Drake *et al.*, 1996.

to support the growth of aerobes, micro-aerophils and anaerobes along an oxygen gradient. Water courses carrying a high load of organic matter may deposit considerable quantities of sediment, especially where slow flowing. Such sediments may be anoxic relatively close to the sediment surface and, depending on sulphate concentration, either sulphate-reducing bacteria or methanogens may be active. Such conditions, however, are more typical of some types of lake (*Feature 6*).

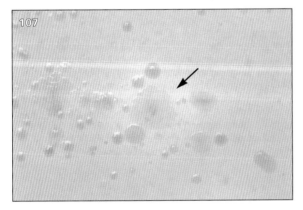

107 A species of *Lysobacter. Lysobacter* (arrowed) is one of the more common gliding bacteria in water. A zone of clearing in the yeast cell substrate may be seen surrounding the colony. The source is particularly rich in *Lysobacter* and other gliding bacteria. (Source: inlet off lowland river; Medium: yeast cell agar; Incubation: room temperature, 3 days.)

108 In running water, *Lysobacter* appears concentrated in discrete areas. Several gliding bacteria have been isolated from the inlet illustrated, which is shallow, slow flowing and contains large quantities of plant debris.

Lakes and ponds

Lakes and ponds are environments that have much in common with running water and, indeed, are often a part of the same water system. A feature of many lakes and ponds is the higher nutrient concentration, stemming from a build-up of deposits and nutrients. This is usually most marked where there is little or no connection with running water systems. Even where strong water flows are present there are likely to be significant areas where water is effectively still and considerable build-up of deposits occurs.

Although the underlying microflora is similar to that of running water, the higher nutrient level of ponds and lakes supports higher numbers of micro-organisms and in some cases a greater variety of types. As the nutrient level increases towards eutrophication the number of types may be reduced by competition. There can be considerable variation between lakes in close geographical proximity (**109**) for which no immediate explanation is available.

As in many other ecosystems, including both fresh and marine waters, the total bacterial population is limited by predation. Studies of a hypertrophic lake (Sommaruga and Psenner, 1995) revealed the permanent presence of grazing-resistant

bacteria. The cells of these comprised filaments, the length of which sometimes exceeded 200 μm (**110, 111**), the accompanying micro-organisms being small cocci. Such bacteria are larger than nano-flagellates and even metazoans, accounting for 45–81% of bacterial biovolume although only 4–16% of bacterial abundance. The large size appears effective as a defence mechanism against predatory ciliates, such as *Cyclidium* spp., but equally may predispose the cells to attack by larger predators such as *Daphnia* spp. Under these circumstances small cocci persist and it seems likely that the cell size of the dominant bacteria varies in response to the physical size of the major predator. In the case of large bacterial cells, it is likely that persistence results from the selective pressure of a high level of predation by nanoflagellates, but absence of larger predators. It is possible, indeed probable, that large cell size is a defence against predation in other habitats.

As with marine environments, bacteriophages play an important role both in limiting bacterial production and short-circuiting carbon flow. A study of a backwater of the Danube River (Mathias and Kirschner, 1995) showed a difference in susceptibility between rod-shaped and 'vibrioid' bacteria and coccoid bacteria, in that the former two

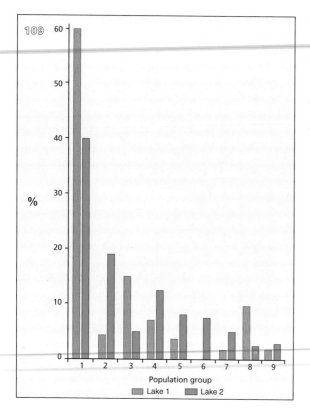

109 A series of small, shallow lakes (maximum depth about 2.0 m) were created as part of a riverside nature reserve designed to reduce the environmental impact of nearby commercial development. Some of the lakes are unconnected but share a small stream as water supply and are separated only by narrow (approx. 2.5 m) embankments bearing hoggin paths. Although total microbial numbers are similar in each lake, the distribution of types of bacteria varies widely.

morphological types were much more susceptible to bacteriophage infection. Virus induced mortality was probably at least 15–30% suggesting that up to 33% of carbon cycles within a microbial loop rather than moving up the food chain (see also page 30).

As well as in sea ice, microbial communities can thrive in the ice and snow cover of high mountain lakes as detailed in *Feature 7*, page 70.

The water column of shallow ponds or shallow areas of larger lakes can contain a particularly wide range of micro-organisms. Such ecosystems appear to be particularly favourable for 'non-motile, Gram-negative, curved bacteria', including *Flectobacillus* (**112**) and *Runella* (**113**). The extent to which Actinomycetes colonize lakes is frequently under-estimated, especially in waters which are relatively

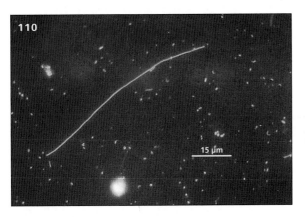

110 A grazing-resistant filamentous bacterium, 75 μm in length, together with a heterotrophic flagellate.

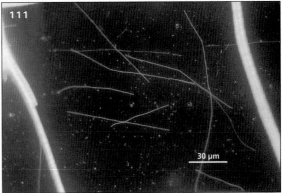

111 An overview of micro-organisms in hypertrophic lake water showing both large filamentous bacteria and small coccal-shaped cells. The filaments in the foreground are of the cyanobacterium *Planktothrix agardhii*. **110** and **111** reproduced with permission from Sommaruga and Psenner, 1995, © 1995 American Society for Microbiology.

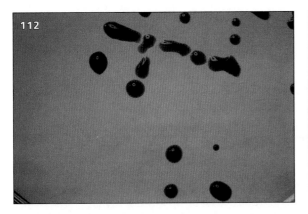

112 A species of *Flectobacillus*. In contrast to *Spirosoma* and *Runella*, a marine species has been isolated having an obligate requirement for the Na⁺ ion. (Source: pond; Medium: MS medium – the addition of 3% NaCl is required for marine species; Incubation: room temperature, 10 days.)

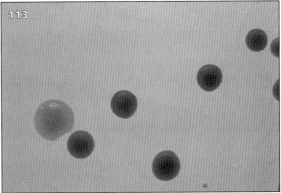

113 *Runella slithiformis*. Distribution was highly variable even over a small sampling area. The yellow colony is of an oxidase-negative, Gram-negative rod. (Source: flooded sand pit – greensand; Medium: MS medium; Incubation: room temperature, 12 days.)

114 Snow cover at Estany Rado during winter.

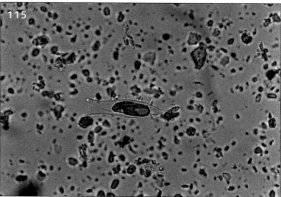

115 The flagellate *Dinobryon cylindricum*, a photosynthetic eukaryote isolated from the Estany Rado snow cover.

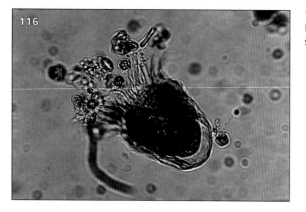

116 The ciliate *Perigostrombidium fallax*, a predator on bacteria in the deeper layers of snow cover during the spring melt.

Feature 7. Rocky mountain high

Colonization of sea ice is a well-known phenomenon (see page 55) but it is often not appreciated that active microbial communities can develop in the ice and snow cover of high mountain lakes. Studies in Estany Rado, Spain (**114**), showed that the winter cover provides a highly dynamic environment, the physical structure and chemical characteristics of which change dramatically as a result of snowfall, periodic melting, freezing and flooding. The main groups of micro-organisms, autotrophic and heterotrophic flagellates (**115**), ciliates (**116**) and bacteria (**117**), respond directly to these physicochemical changes with resulting effects on biomass and species composition. There appear to be two distinct phases which have been attributed to different mechanisms of colonization and growth. From January to mid-April (the phase of growth of the cover), microbial assemblages are related to their planktonic counterparts. During this period, colonization is primarily due to plankton derived from the lake water which floods the cover. As light fails to reach the slush layers, the plankton is primarily dependent on bacterivory for growth.

From mid-April to the spring melt in June (ablation), the physicochemical nature of the environment is strongly influenced by the large amount of water (**118**) from the melting of the snowpack. The water is a source of both nutrients and micro-organisms, while the light availability permits the growth of algae and an associated food web. In deeper layers the food web is based on bacteria, providing prey for large ciliates, which are probably derived from littoral or watershed soils. **114–116** and **118** reproduced by courtesy of Dr M. Felip, University of Barcelona, Spain; **117** reproduced with permission from Felip *et al.*, 1995, ©1995 American Society for Microbiology.

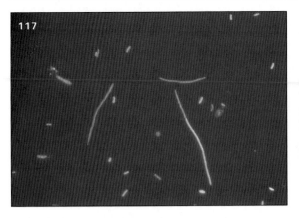

117 Bacteria, viewed by epifluorescent microscopy after staining with DAPI, from the deeper layers of snow cover during the spring melt. Note the presence of elongated cells; cf. **110**, **111**.

118 Water from melting snow cover at Estany Rado during spring (May). The scale of the melt is apparent from the quantity of water in the foreground.

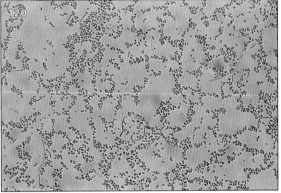

119 *Legionella pneumophila*. Methods for isolation from natural environments are often time consuming, the most suitable method varying according to the source of the sample (Kusnetsov *et al.*, 1994). Colonies on the non-selective medium illustrated may also have a ground glass appearance. (Source: thermally polluted, disused canal; Medium: buffered charcoal–yeast extract agar supplemented with ∝-ketoglutarate – Edelstein, 1981; Incubation: 35°C (95°F), 7 days.)

120 *Legionella pneumophila*. Cells are highly pleomorphic, varying from short cocco-bacilli to long filaments, though filaments are not present during growth in animal tissue. Staining is often weak and may be irregular. (Source: as for **119**; Microscopy: Gram stain, × 1,200.)

nutrient rich. Actinomycetes are important in the decomposition of organic compounds such as chitin and cellulose. The dominant genus is often *Micromonospora* with *Streptomyces* second in importance (Jiang and Xu, 1996).

Species of the human pathogen *Legionella* (**119**, **120**) are common in natural waters, especially those of high nutrient content and thermally polluted. Waters of this type appear to serve as a reservoir of

Legionella from which the bacterium enters and colonizes water cooling systems, humidifiers, and the like. Some of these provide a very favourable environment for *Legionella* in terms of temperature, level of aeration, organic nutrients and iron concentration. In either natural or artificial systems the organism may develop as a member of a biofilm but there is also an association between *L. pneumophila* and the free-living amoeba

Acanthamoeba polyphaga (**121**) (Kilvington and Price, 1990). This involves the intracellular growth of *L. pneumophila* inside amoebic trophozoites. From a public health viewpoint this association is of concern because it enables *L. pneumophila* to survive chlorination of cooling water and similar systems. Aerial distribution inside amoebae, along with fragments of biofilm and large droplets, may also represent a means by which the bacterium bypasses defence mechanisms in the lung. From a broader environmental viewpoint, intracellular growth is probably beneficial to *L. pneumophila* in preventing predation by other protozoa, ensuring a supply of nutrients, including iron, and protecting against UV irradiation to which *Legionella* is highly sensitive.

The reduced O_2 concentration in the upper layer of sediments is favourable for growth of *Serpens* and also for true micro-aerophiles such as *Spirillum*.

Spirillum and similar bacteria may also be found below the surface scum of stagnant water where favourable micro-aerophilic conditions are created by the surface growth of aerobic bacteria. In heavily polluted waters, rich in organic matter, the surface film may include the enigmatic *Lampropedia hyalina* (**122**) although the real habitat of this bacterium is not known.

A number of the micro-aerophilic and anaerobic bacteria of the sediment of lakes and ponds have been shown to exhibit magnetotactic behaviour. These bacteria, which are motile and morphologically diverse, contain high quantities of iron (c.0.4% of the dry weight of the cell) as magnetite (Fe_3O_4), the iron sulphide greisite (Fe_3S_4) or a combination of greisite with iron pyrites (FeS_2) which is incorporated as specialist sensing organelles, the magnetosomes (**123**). These function

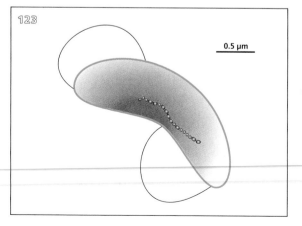

121 Diagrammatic representation of *L. pneumophila* inside an *Acanthamoeba* cell.

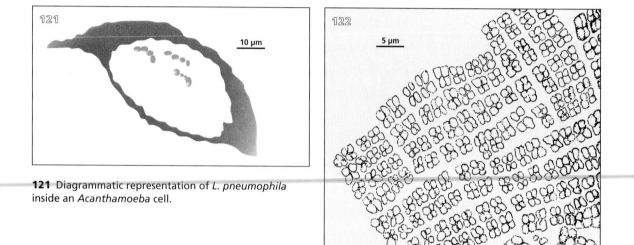

122 *Lampropedia hyalina* showing the characteristic 'window pane' morphology.

123 A general view of a cell of a magnetotactic bacterium showing the position of magnetosomes. Magnetotactic bacteria in the northern hemisphere are almost entirely north seeking while those in the southern hemisphere are almost entirely south-seeking. Populations collected at the geomagnetic equator, where the magnetic field has no vertical component, are an equal mixture of north- and south-seeking cells.

by orientating cells in the unfavourable O_2-containing water column so that their motility results in a return to the favourable conditions of the sediment. There has been a recent description of an organism which contains both magnetite and greisite magnetosomes (**124–126**).

Considerable anaerobic activity occurs in the sediment and can involve both sulphate-reduction and methanogenesis (see page 75). In shallow lakes of high organic matter content, anaerobic condi-tions can extend throughout the water column with the exception of the water just below the air interface which is occupied by cyanobacteria and eukaryotic algae. A neuston may be formed by *Euglena sanguinea* which contains photoprotective pigments. These migrate within the cell according to light intensity and the water colour may also change. Where significant quantities of H_2S are produced in the underlying sediment (cf. marine sediments), purple and green sulphur bacteria (**127**)

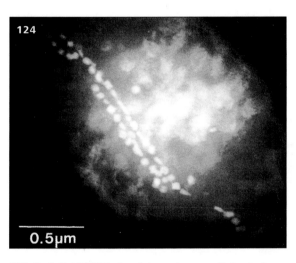

124 Dark field STEM of rod-shaped, magnetic bacteria showing two chains of magnetosomes. Magnetite magnetosomes are arrow shaped while greisite magnetosomes are rectangular. Reproduced by courtesy of Dr D. Bazylinski, Iowa State University, Ames.

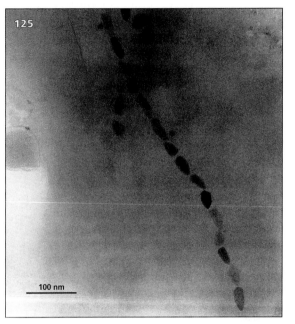

125 TEM showing arrangement of magnetosomes Reproduced by courtesy of Drs D. Bazylinski and B.R. Hayward, Iowa State University, Ames.

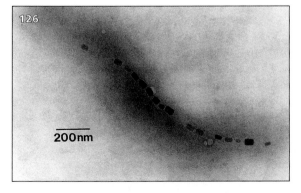

126 A magnetotactic micro-organism. Despite the large gap between some magnetosomes, the consistent orientation suggests co-organization by a chain assembly process. Reproduced by courtesy of Drs D. Bazylinski and F.C. Meldrum, from Bazylinski *et al.*, 1995, © 1995 American Society for Microbiology.

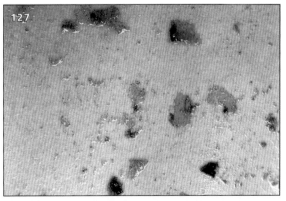

127 Green sulphur bacteria. Growth below the oxic surface layer of a predominantly anoxic lake consisted of flakes containing green sulphur bacteria and an unknown non-photosynthetic, Gram-negative bacterium. The bacteria were embedded in a waxy matrix.

develop in a layer directly below the oxic upper zone. Development of anaerobic conditions in shallow lakes can be a particular problem in hot weather when oxygen is less soluble in water and microbial activity high. Aeration of water is often required to prevent the death of fish and recreational lakes, marinas, and the like, can easily lose all amenity value. Anaerobic activity in the sediment extends into the water column and includes outgrowth of clostridial spores (**128**). Under these conditions toxin production by *Cl. botulinum* has caused the death of aquatic birds, including swans and, allegedly, bathing dogs (**129, 130**).

Deep lakes can exhibit stratification of the water column into aerobic and anaerobic zones (**131**). This results from the differing density of water at different temperatures, density being a maximum at 4°C (39.2°F). Solar heating produces a layer of warm water of low density, the epilimnion, which floats on the colder, denser layer, the hypolimnion. The two layers are separated by the thermocline (chemocline) at a depth of 20–30 m. The upper layer of water is aerobic although where geographical conditions limit mixing an oxygen gradient may be created extending to the thermocline. Conditions are anaerobic below the thermocline with considerable anaerobic activity occurring in the sediment.

Stratified lakes are of two types. In meromictic lakes, stratification is permanent and persists independent of seasonal changes. Meromictic lakes occur most commonly in tropical and subtropical zones but can also occur in temperate climates. Stratified marine basins (see page 34) are of this type. In holomictic lakes, stratification is seasonal. The epilimnion is established when ambient temperatures rise in the spring and early summer. An algal and

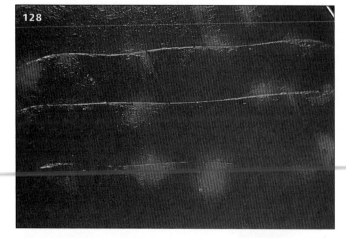

128 A putrefactive anaerobe (*Clostridium* sp.). The isolate was phenotypically similar to proteolytic strains of *Cl. botulinum*. (Source: anoxic lake sediment; Medium: blood agar; Incubation: 37°C (98.6°F), 24 hours, 95% H_2:5% CO_2.)

129 *Clostridium botulinum* grows rapidly under suitable conditions. In lakes, corpses of dead birds and animals may be a focal point for growth and toxigenesis.

130 The environment may become sufficiently contaminated by *Clostridium botulinum* to present a severe risk to aquatic birds, including swans.

cyanobacterial bloom may occur at this stage due to a combination of higher temperature and an elevated nutrient concentration following mixing of the water layers the previous winter. Stratification usually remains stable over the summer, until the temperature of the epilimnion falls during the autumn. At some stage, usually during early winter, the temperature of the epilimnion falls below that of the hypolimnion and stratification becomes unstable. Mixing of the layers occurs, often initially promoted by strong winds. This results in some oxygenation of lower layers and distribution of nutrients from the lower layers throughout the water column. Inverse stratification can occur in the winter when deep, high density water at 4°C (39.2°F) is overlaid by colder, less dense water and ice. Mixing of the layers when surface water temperatures rise in late winter leads to nutrient distribution and contributes to the spring bloom.

Stratification and resulting chemical gradients leads to intense microbiological activity at different depths in the lake (**132**). Organic matter tends to accumulate in the sediment, where anaerobic degradation occurs. The ultimate fate of organic compounds largely depends on the relative importance of sulphate-reducing and methanogenic bacteria. Sulphate-reducing bacteria normally dominate, even where the quantity of organic deposition is such that adequate H_2 is available, permitting a degree of coexistence between the two groups. Sulphate levels in freshwater tend to be much lower than in seawater and can become limiting. Under these conditions production of both H_2S and CH_4 occurs in the sediment, CH_4 being the major end-product. These gases, together with CO_2, enter the water column where oxidation of CH_4 by methophils leads to anaerobic conditions. Methane is only sparingly soluble in water and significant quantities escape as gas bubbles.

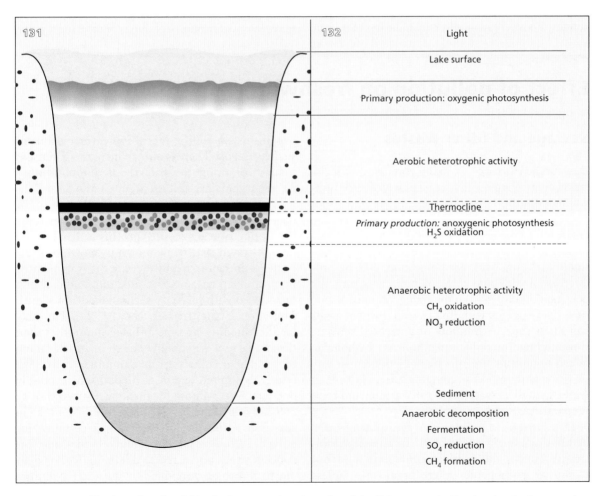

131, 132 Stratification of, and activity of micro-organisms in, a deep lake. Primary production involves both oxygenic and anoxygenic photosynthesis in distinct zones. This situation occurs only in lakes of high nutrient content.

In the presence of H_2S, anoxygenic phototrophs (purple and green sulphur bacteria) develop in a band just below the thermocline where the spectral quality of light penetrating and the chemical environment is optimal. Anaerobic, heterotrophic micro-organisms, capable of withstanding high H_2S concentrations are also present in large numbers in the anaerobic hypolimnion, while both dissimilatory nitrate reduction and denitrification also occur. Sulphate-reducing bacteria are active, utilizing SO_4^{2-} produced by purple sulphur bacteria.

In the epilimnion, primary production involves oxygenic photosynthesis at the air–water interface. This usually involves both eukaryotic algae and cyanobacteria but when H_2S is present cyanobacteria such as *Oscillatoria* are favoured. A high level of heterotrophic activity occurs below the surface layer. Heterotrophic activity increases with higher temperatures, but there is a reduction in diversity. This is a general trend in freshwater and reflects the dominance of strains of high heterotrophic activity. The decrease in diversity indicates that some members of the water population are stressed by even relatively small rises in water temperature.

In northern Europe and North America a number of lakes have undergone acidification over the past decades. This stems from accelerated acidification of soil in the catchment area. Acidification is a natural process in soils with a hard underlying geology, exacerbated in some cases by increased afforestation and acid precipitation. Studies of the effects on the microflora of the water column and the sediment have produced conflicting results. Acidification has been variously reported to reduce the viable heterotrophic count by a log cycle or to have no effect. In a study of freshwater lochs in south-west Scotland, there was also little effect on the diversity of the heterotrophic population, in the pH range 3.8–6.5, and only one organism, a species of *Cytophaga*, was present at lower numbers at reduced pH (Rattray and Logan, 1993). Acidification has a pronounced effect on protozoa and on larger animals including fish. Attempts have been made, especially in Scandinavia, to reverse acidification by liming the catchment areas.

Effect of pollution on freshwater environments

Sewage and other wastes

Despite the existence, in many countries, of legislation controlling the discharge of waste into freshwater, pollution from sewage outlets and other sources can still cause considerable problems. The environmental consequences of discharges of large quantities of untreated sewage are well known, resulting in large-scale eutrophication, massive production of H_2S and CH_4 in sediments and the depletion of oxygen from the water column. Despite this, river systems with low overall nutrient loads can often cope with inputs of untreated sewage, through heterotrophic metabolism. Pathogenic micro-organisms are of obvious concern in any situation with respect to contamination of drinking water, as well as water used for washing and food preparation.

In industrialized countries sewage is often treated before discharge (see page 80). Treatment significantly reduces the nutrient input but does not eliminate pathogenic micro-organisms unless treated with high levels of chlorine. Readily observable effects (to the practising microbiologist) include an increase in numbers of enteric micro-organisms, including faecal indicator organisms such as *E. coli*. There is also an increase in nutrient load, promoting the growth of autochthonous micro-organisms, and an input of allochthonous micro-organisms. An increase in the density of the bacterial population may be expected to increase the population of grazing protozoa. It has been demonstrated that while a high proportion of the bacteria present in freshwater are easily eliminated by protozoan grazing, the allochthonous bacteria introduced with sewage are eliminated at a significantly lower rate (Irriberri *et al.*, 1994). This may be attributed to the type of bacterium rather than the source and is probably due to protozoa failing to recognize signals from unfamiliar cells. The relatively limited extent of predation is probably one factor accounting for the longer survival of *E. coli* and other enteric bacteria in freshwater than in seawater. Other factors are almost certainly involved, including lack of salinity, higher nutrient concentration and a reduced lethal effect of light due to absorbing compounds in the water. Numbers of bacteriophages and *Bd. bacteriovorus* increase in the vicinity of sewage outlets, probably as a consequence of the higher host density.

Sewage has been considered to be a potential source of antibiotic-resistant bacteria. The evidence for the occurrence of genetic transfer in natural waters is conflicting but it has been suggested that the spread of antibiotic resistance among environmental bacteria poses risks to public health. In some surveys the level of resistance amongst enteric bacteria originating in sewage has been very high, resistance patterns reflecting the prescription of antibiotics for medical use (Morinigo *et al.*, 1990).

Other urban waste water, including surface runoff and water used for domestic purposes, is of relatively low nutrient concentration and contributes relatively few allochthonous bacteria, although *Aer. hydrophila* may be present in large numbers. Both this waste water and sewage can be a significant source of inorganic pollution, including phosphates, and can be a major contributor to algal and cyanobacterial blooms.

Pollution from industrial sources

Fresh water may receive a vast variety of industrial pollutants. Industrialization of an economy seems invariably to involve a stage at which environmental pollution goes virtually unchecked. Even in mature industrial economies, where environmental concerns are recognized, a considerable level of pollution may occur due to seepage from old sites, accidental spillage or illegal disposal. A vast number of pollutants are potentially involved, ranging from the relatively innocuous (in human terms) wastes from agro-industrial operations to highly toxic chemicals and radioactive materials. The effect of pollution depends on the nature and concentration of the pollutants and on whether pollution occurred on a single occasion or was continuous. Waste of high organic content, such as that from potato processing, may be expected to cause eutrophication with selection for heterotrophic micro-organisms able to utilize the nutrients most efficiently. Pollution with toxic chemicals generally leads to a loss of diversity and the emergence of dominant, resistant populations. Diversity may increase as resistant sub-populations of sensitive micro-organisms develop and micro-organisms capable of metabolizing the pollutant increase in number. Microbial consortia in sediments are often of considerable importance in metabolizing recalcitrant pollutants. In some cases heavy and continuing pollution has a very dramatic effect on diversity, which continues over an extended period (**133, 134**).

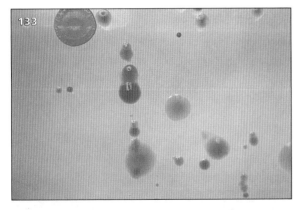

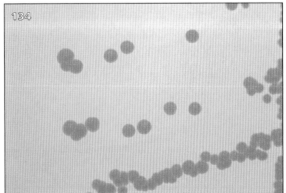

133 The effect of heavy metal pollution on the microflora of a stream. A limestone stream in a remote area has a typically mixed microflora, predominantly Gram-negative rods. Compare with **134**.

134 Below the point of pollution with surface water emerging from a defunct lead mine only a single, yellow-pigmented micro-organism can be recovered. The organism is distinct from dominant strains isolated above the point of pollution. (Source: limestone stream, above and below point of lead pollution; Medium: quarter-strength nutrient agar; Incubation: 20°C (68°F), 7 days.)

Relationship between freshwater micro-organisms and aquatic plants

Aquatic plants are of much greater importance in freshwater than in marine environments. The general relationship reflects that of micro-organisms with terrestrial plants. Colonization of roots and their environs is analogous to the rhizosphere and bacteria present can be either beneficial or deleterious to the plant. A similar, but usually less diverse, microflora develops on the roots of plants grown in soilless nutrient solutions.

Colonization of the leaves and shoots of aquatic plants appears variable. In some cases a biofilm develops, members of which are similar to micro-organisms in the water column and colonizing inert material. In other cases, the epiphytic microflora is distinct from that in surrounding water. Reed beds fringing an artificial lake (**135**) show a distinctive pattern, in which submerged stems of live plants are colonized by a single dominant Gram-negative bacterium (**136**). This microflora differs from the highly varied microflora in the water of the reed bed (**137**) and from the water in the body of the lake (**138**). The epithytic microflora on reed stems above the water level was also distinct from that below (**139, 140**).

The floating aquatic fern *Azolla* forms an economically important symbiotic relationship with the heterocystous cyanobacterium *Anabaena azollae* (**141**). *Anabaena* is harboured within mucilage-containing cavities on the leaves of *Azolla* (**142**). This symbiosis is of very considerable importance in supplying nitrogen to rice paddies in southeast Asia, *Azolla* being co-cultivated with rice. Dense growth of *Azolla* is also beneficial in reducing weed growth and is harvested for use as green manure or poultry feed. *Anabaena* is commercially available as an inoculant and the use of the *Azolla* symbiosis as means of enhancing rice cultivation has been extended to other growing areas.

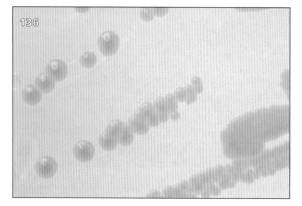

135 Reed beds, planted during the creation of a nature reserve. Reeds carry oxygen to the roots creating a complex system of microhabitats housing a large microbial population.

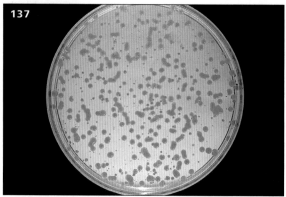

136 A Gram-negative, oxidase-positive, rod-shaped bacterium dominant on the submerged surface of reeds. (Medium: quarter-strength nutrient agar; Incubation: 22°C (71.6°F), 5 days.)

137 Microflora of water in a reed bed. A range of Gram-negative rods (oxidase positive and oxidase negative) and smaller numbers of Gram-positive cocci were dominant. (Medium: quarter-strength nutrient agar; Incubation: 22°C (71.6°F), 5 days.)

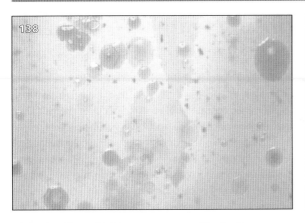

138 Microflora of lake water near reed beds. Gram-negative rods, including yellow-pigmented and swarming strains, are dominant. (Medium: quarter-strength nutrient agar; Incubation: 22°C (71.6°F), 5 days.)

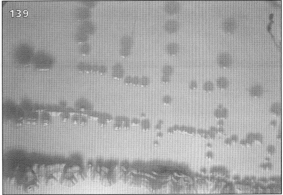

139 A species of *Erwinia*, one of the two dominant epithytic bacteria of reed stems above the water level. (Medium: quarter-strength nutrient agar; Incubation: 22°C (71.6°F), 5 days.)

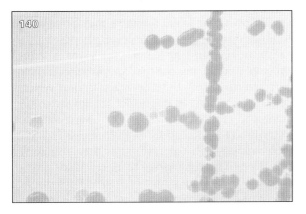

140 A species of *Pseudomonas*, one of the two dominant epithytic bacteria of reed stems above the water level. (Medium: quarter-strength nutrient agar; Incubation: 22°C (71.6°F), 5 days.)

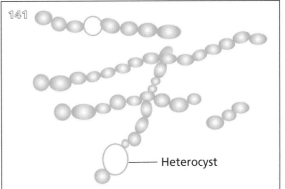

Heterocyst

141 Diagrammatic representation of filaments of *Anabaena* (x 700).

142 The *Azolla* symbiosis. In the immature leaf, mucilage-containing cavities are open, but close with maturity. These cavities harbour the filaments of *Anabaena*.

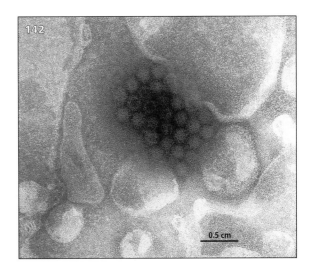

0.5 cm

Treatment of sewage and drinking water

Sewage treatment

Mineralization of organic matter in natural waters plays an important role in the cycles of matter. The growth of the human population and its concentration in relatively small geographical areas means that natural processes are no longer adequate to dispose of waste material. One of the prime roles of sewage collection (**143**) and treatment is the intensive operation of substantial parts of the cycles of matter under relatively controlled conditions. A variety of approaches may be taken to sewage treatment but the principle is the same in each case. Solids which form sediment relatively readily are removed by various grit and sedimentation chambers. The liquid then undergoes biological purification that mimics mineralization in the natural environment. For many years the most widely used means was the percolating filter. The most common form involved rotating arms spraying the liquid onto a bed of loosely packed rocks, or clinker, through which it percolates by gravity. Percolating filters are still widely used but have been replaced in many plants by equipment using active aeration. In the activated sludge system, air is blown upwards through a tank of liquid effluent (**144**) while variants include 'racetracks' comprising a system of circular channels through which effluent is pumped under conditions designed to promote aeration. In percolating filters the degradation of organic matter is carried out by communities of micro-organisms in biofilms which develop on the surfaces of the filter, while in activated sludge plants and similar systems, flocs are formed which are composed of actively metabolizing bacteria. The role of micro-organisms is similar in each case and while the process is primarily aerobic, chemical gradients across both the film and floc are such that anaerobic activity also takes place. Treated liquid is discharged into rivers or the ocean.

Sediment removed at the preliminary screening stage, together with sedimented flocs produced by activated sludge processes are passed to anaerobic digestion chambers. Organic input provides sufficient H_2 for both sulphate-reducing bacteria and methanogens. Methane production is very considerable and is commonly collected, compressed and used as a fuel for electrical generators. The sludge remaining after anaerobic digestion is dried, composted and used as fertilizer, or burnt. In the former case there has been concern over the possible survival of pathogens and heavy metal levels.

Although well designed and correctly managed sewage disposal plants produce a highly mineralized liquid effluent, products including nitrate, phosphate, and ammonium are present and can be significant contributors to eutrophication. Further chemical treatments are available to remove these ions or, alternatively, the nitrogen may be removed by denitrification following application to soil or used for fertilization of forests.

For many years it was thought that sewage treatment effectively removed enteric pathogens but it is now known that while numbers may be reduced, elimination is not possible, even though chlorination of the effluent leaving the plant is highly effective in reducing the load of pathogens, including enteroviruses (**145**). Sewage disposal systems do, however, play an important role in separating human waste from sources of drinking water, or water used for recreation. Both historically and in recent times, such separation has been a factor in reduction of morbidity due to enteric disease.

Sewage treatment plants often develop a distinct microflora, in which prosthecate and/or filamentous bacteria are commonly favoured. An example is that of *Hyphomicrobium*, a Gram-negative methylotrophic bacterium that reproduces by budding at the tips of the cellular prosthecae, the hyphae (**146**). *Hyphomicrobium* is one of a group of morphologically related bacteria including *Hirschia*, *Hyphomonas*, *Pedomicrobium* and *Rhodomicrobium*, but differs physiologically in its ability to utilize Cl compounds. *Hyphomicrobium* is found in a diversity of aquatic habitats as well as in soil and its ecological importance may well be under-estimated. The organism attaches to flocs formed in the activated sludge sewage treatment process (**147**) and is notable for its genetic diversity which leads to different, and distinct, populations of *Hyphomicrobium* being present in different parts of sewage treatment plants (Holm *et al*., 1996).

In the UK, however, there is no statutory requirement to remove pathogens. The basis of sewage treatment is the reduction of Biological Oxygen Demand (BOD) to satisfactory, legally defined, levels.

143 Efficient sewage disposal is taken very much for granted in the western world and municipal engineering is hardly a fashionable subject area. In earlier times, however, the water supply and sewage disposal systems were justifiably viewed alongside railways as triumphs of engineering science and a considerable source of civic pride. Sir Joseph Bazalgette was Engineer to the Metropolitan Board of Works and responsible for the London, UK, sewage and drainage system.

144 An activated sludge plant. Success of the process depends on the microbial community forming flocs, which rapidly settle leaving clear effluent. The process is disrupted by 'bulking', when filamentous organisms extend from the floc into the bulk solution interfering with clarification. Few of the bulking organisms have been isolated in pure culture and the classification of some, such as *Michothrix parvicella*, has been doubted.

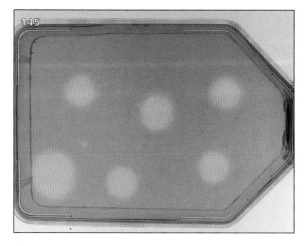

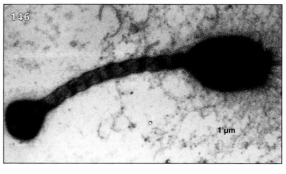

146 *Hyphomicrobium* sp. from a sewage treatment plant, showing helically twisted hypha.

145 Poliovirus from raw sewage forming plaques in monkey kidney cell monolayer.

147 An activated sludge floc with attached cells of *Hyphomicrobium*. **146** and **147** reproduced with permission from Holm *et al.*, 1996, © 1996 American Society for Microbiology.

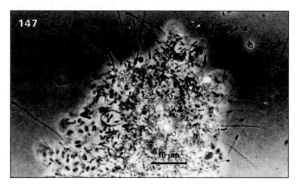

Drinking water treatment

A considerable part of drinking water treatment involves the removal of suspended and dissolved organic matter. For many years slow sand filters were used which involved the development of a microbial biofilm on the filter bed. This process is now considered uneconomical on a large scale and there is now a greater dependence on purely physical filtration (fast sand filtration) and physicochemical methods such as flocculation. In each case final disinfection of the water is required, the most commonly used treatment being chlorination. Levels of chlorination are usually adjusted to allow a level of residual chlorine to offer some protection against contamination in the distribution system. In many countries public water supplies have an extremely good safety record and microbial numbers are very low at point of consumption if drawn direct from the piped supply. Most concern over the safety of water has involved the presence of ions, especially nitrite and recalcitrant agricultural chemicals, especially pesticides. Microbiological problems have, however, emerged involving the protozoan parasites *Giardia lamblia* (**148**) and *Cryptosporidium* (**149–152**). In a

number of cases, illness has been attributed to defective filtration, oocysts not removed at this stage being resistant to chlorination. From an ecological viewpoint it is instructive to consider these protozoa in the context of their widespread distribution in the environment. *Giardia*, for example, is present in even pristine stream water and a cause of 'trekkers' trots'. Co-infection with *Campylobacter* is common in this situation and the two organisms may share a reservoir amongst small animals. *Cryptosporidium* is extremely widespread in the environment and routes of transmission to man are complex.

In the UK and other industrialized countries concern, usually unjustified, over the quality of piped water has led to a large increase in the consumption of bottled spring and well water. Despite the 'sparkling pure' image promoted for this water, microbial numbers (in non-carbonated water) can be high (**153**). In the US problems have also been caused by use of well water for ice making (**154**). A further development has been increased sales of home water filters. These devices may be a source of micro-organisms and also become colonized by aquatic micro-organisms shedding large numbers into the water (**155**).

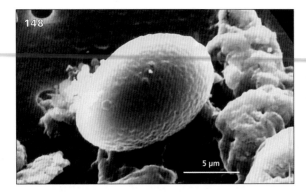

148 *Giardia lamblia*. Scanning electron micrograph of trophozoites.

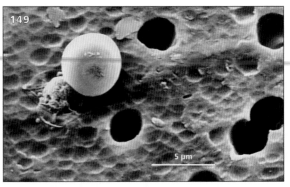

149 *Cryptosporidium parvum*. Scanning electron micrograph of oocysts.

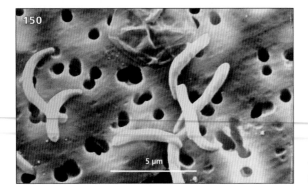

150 *Cryptosporidium parvum*. Scanning electron micrograph of sporozoites.

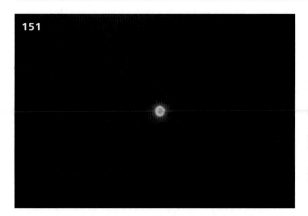

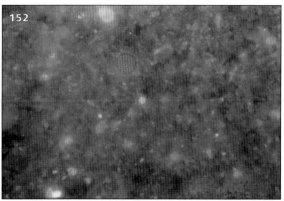

151 Detection of *Cryptosporidium* oocysts in water. Detection involves the filtration of large volumes of water followed by fluorescent microscopy. Oocysts are easily detected in isolation.

152 Detection of *Cryptosporidium* oocysts in water. The examination of actual samples is tedious and time consuming, with a possibility of overlooking *Cryptosporidium* against the background material. **144, 145, 148–152** reproduced by courtesy of Thames Water Utilities Ltd, Reading.

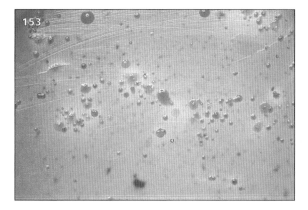

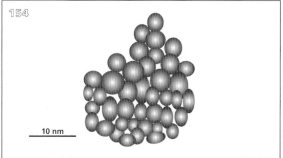

154 Ice and viral diarrhoea. Diagrammatic representation of a small, round, structured virus (i.e. SRSV). A virus of this type was responsible for more than 950 known cases of illness in Pennsylvania and Delaware, USA. The source was traced to ice produced from wells contaminated with river water during flooding. The well water, which was not treated before use, was probably contaminated by sewage.

153 The microflora of uncarbonated bottled spring water. Numbers present were in the order of 10^4 cfu ml^{-1}. The flora largely comprised Gram-negative rods and was similar to that in unpolluted streams. Some members of the microflora may be opportunistic pathogens. (Medium: quarter-strength nutrient agar; Incubation: 22°C (71.6°F), 6 days.)

155 Shedding of bacteria into water by a domestic filter. The interior of the filter was colonized by a biofilm comprising a species of *Pseudomonas* (large colonies) and a Gram-positive, irregular, rod-shaped bacterium of unknown identity (small colony). Although psychrotrophic rather than psychrophilic, this organism developed visible colonies only after extended incubation at 5°C (41°F). (Medium: quarter-strength nutrient agar; Incubation: 22°C (71.6°F), 4 days; 5°C (41°F), 15 days.)

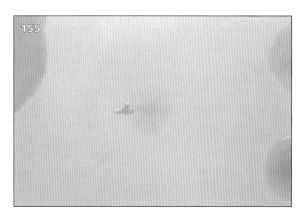

REFERENCES

Baumann, P., Furniss, A.L. and Lee, J.V. (1984) Vibrio, in *Bergey's Manual of Systematic Bacteriology* (eds. Krieg, N.R. and Holt, J.G.). Williams & Wilkins, Baltimore.

Bazylinski, D.A., Frankel, R.B., Heywood, B.R. *et al.* (1995) Controlled mineralization of magnetite (Fe_3O_4) and griesite (Fe_3S_4) in a magnetotactic bacterium, *Applied and Environmental Bacteriology*, **61**, 3232–9.

Bebout, B.M. and Garcia-Pichel, F. (1995) UV-B-induced vertical migrations of Cyanobacteria in a microbial mat, *Applied and Environmental Microbiology*, **61**, 4215–19.

Bermingham, S., Dewes, F.M. and Maltby, L. (1995) Development of a monoclonal antibody-based immunoassay for the detection and quantification of *Anguillospora longissima* colonizing leaf material, *Applied and Environmental Microbiology*, **61**, 2602–13.

Bordalo, A.A. (1993) Effects of salinity on bacterioplankton: field and microcosm experiments, *Journal of Applied Bacteriology*, **75**, 393–8.

Carlton, R.G. and Richardson, L.L. (1995) Oxygen and sulfide dynamics in a horizontally migrating cyanobacterial mat: Black band disease of corals, *FEMS Microbiology Ecology*, **18**, 155–62.

Cavanaugh, C.M., Wirsen, C.O. and Jannasch, H.W. (1992) Evidence for methylotrophic symbionts in a hydrothermal vent mussel (Bivalvia: Mustelidae) from the mid-Atlantic ridge, *Applied and Environmental Microbiology*, **58**, 3799–803.

DePaola, A., Capers, G.S. and Alexander, D. (1994) Densities of *Vibrio vulnificus* in the intestines of fish from the US Gulf Coast, *Applied and Environmental Microbiology*, **60**, 984–8.

Drake, H.L., Aumen, N.G., Kuhner, C. *et al.* (1996) Juxtaposition of large human populations with sensitive ecosystems leads to a high level of anthropogenic stress, *Applied and Environmental Microbiology*, **62**, 486–93.

Edelstein, P.H. (1981) Improved semielective medium for isolation of *Legionella pneumophila* from contaminated clinical or environmental specimens, *Journal of Clinical Microbiology*, **14**, 298–302.

Fay, P. (1992) Oxygen relations of nitrogen fixation in cyanobacteria, *Microbiological Reviews*, **56**, 340–73.

Felip, M., Sattler, B., Psenner, R. and Catalan, J. (1995) Highly active microbial communities in the ice and snow cover of high mountain lakes, *Applied and Environmental Microbiology*, **61**, 2394–401.

Fernandez-Alvarez, R.M., Carballo-Cuervo, M., De La Rosa-Jorge, C. and Rodriguez-De Lecea, J. (1991) The influence of agricultural run-off on bacterial populations in a river, *Journal of Applied Bacteriology*, **70**, 437–42.

Fredriksson, C. and Bergman, B. (1995) Nitrogenase quantity varies diurnally in a subset of cells within colonies of the non-heterocystous cyanobacteria *Trichodesmium* spp., *Microbiology*, **141**, 2471–8.

Gautier, M.J., Flatau, G.N., Clement, R.L. and Munro, P.M. (1992) Sensitivity of Escherichia coli cells to seawater closely depends on their growth stage, *Journal of Applied Bacteriology*, **73**, 257–262.

Gosinski, J.J., Irgen, R.L. and Staley, J.T. (1993) Vertical distribution of bacteria in arctic sea ice, *FEMS Microbiology Ecology*, **102**, 85–90.

Hansen, L.S. and Blackburn, T.H. (1992) Mineralization budgets in sediment microcosms: effect of the infauna and anoxic conditions, *FEMS Microbiology Ecology*, **102**, 33–43.

Hennes, K.P., Suttle, C.A. and Cham, A.M. (1995) Fluorescently labelled virus probes show that natural virus populations can control the structure of marine microbial communities, *Applied and Environmental Microbiology*, **61**, 3623–7.

Herbert, R.A. (1992) The application of microelectrodes in microbial ecology, *Journal of Applied Bacteriology Supplement*, **73**, 164–72.

Hespell, R.B. (1984) *Serpens*, *Bergey's Manual of Systematic Bacteriology*, Vol. 1 (eds. N.R. Krieg and J.G. Holt), pp. 373–5, Williams and Wilkins, Baltimore.

Hoaki, T. (1995) Dense community of hyperthermophilic sulphur-dependent heterotrophs in a geothermally heated shallow submarine biotope near Kodakara-Jima Island, Kagoshima, Japan, *Applied and Environmental Microbiology*, **61**, 1931–7.

Holm, N.C., Gliesche, C.G. and Hirsch, P. (1996) Diversity and structure of *Hyphomicrobium* populations in a sewage treatment plant and its adjacent receiving lake, *Applied and Environmental Microbiology*, **62**, 522–8.

Irriberri, J., Ayo, B., Artolozaga, I. *et al.* (1994) Grazing on allochthonous *vs* autochthonous bacteria in river water, *Letters in Applied Microbiology*, **18**, 12–14.

Isaksen, M.F. and Jorgensen, B.B. (1996) Adaptation of psychrophilic and psychrotrophic sulphate-reducing bacteria to permanently cold marine environments, *Applied and Environmental Microbiology*, **62**, 408–14.

Jiang, C.-L. and Xu, L.H. (1996) Diversity of aquatic actinomyces in lakes of the middle plateau, Yunan, China, *Applied and Environmental Microbiology*, **62**, 478–483.

Jorgensen, B.B. (1982) Ecology of the bacteria of the sulphur cycle with special reference to anoxic–oxic interface environments, *Philosophical Transactions of the Royal Society, London, B*, **298**, 543–61.

Jorgensen, B.B. (1994) Sulfate reduction and thiosulphate transformations in a cyanobacterial mat during a diel oxygen cycle, *FEMS Microbiology Letters*, **13**, 302–12.

Kilvington, S. and Price, J. (1990) Survival of Legionella pneumophila within cysts of Acanthamoeba polyphaga following chlorine exposure, *Journal of Applied Bacteriology*, **68**, 519–527.

Konhauser, K.O., Schultze-Lam, S., Ferris, F.G. *et al.* (1994) Mineral precipitation by epilithic biofilms in the Speed River, Ontario, Canada, *Applied and Environmental Microbiology*, **60**, 549–53.

Kuehr, C.S.W., Katarszky, P. and Brunton, M. (1992) Contaminated marine wounds - the risk of acquiring acute bacterial infection from marine recreational beaches, *Journal of Applied Bacteriology*, **73**, 412–420.

Kusnetsov, J.M., Jousimies-Somer, H.R., Nevalainen, A.I. and Martikainen, P.J. (1994) Isolation of Legionella from water samples using various culture methods, *Journal of Applied Bacteriology*, **76**, 155–162.

Lillebaek, R. (1995) Application of antisera raised against sulfate-reducing bacteria for indirect immunofluorescence of immunoreactive bacteria in sediment from the German Baltic Sea, *Applied and Environmental Microbiology*, **61**, 3436–42.

Lovley, D.R. (1991) Dissimilatory Fe(III) and Mn(IV) reduction, *Microbial Reviews*, **55**, 259–87.

Malkii, E. (1988) Biological treatment of groundwater in floating basins with floating filters. 1. Test arrangements and general results, *Water Science and Technology*, **20**, 179–84.

Marteinsson, V.T., Birrien, J.-L., Kristajansson, J.K. and Priean, D. (1995) First isolation of thermophilic aerobic non-sporulating heterotrophic bacteria from deep-sea hydrothermal vents, *FEMS Microbiology Ecology*, **18**, 163–74.

Mathias, C.B. and Kirschner, A.K.T. (1995) Seasonal variations of virus abundance and viral control of the bacterial population in a backwater system of the Danube River, *Applied and Environmental Microbiology*, **61**, 3734–40.

Meighen, E.A. (1992) Bioluminescence, bacterial, *Encyclopedia of Microbiology*, Vol. 1 (ed. J. Lederberg), pp 309–19, Academic Press, New York.

Moran, M.A., Rutherford, L.T. and Hodson, R.E. (1995) Evidence for indigenous *Streptomyces* populations in a marine environment determined with a 16S rRNA probe, *Applied and Environmental Microbiology*, **61**, 3695–700.

Morinigo, M.A., Cornax, R., Castro, D. (1990) Antibiotic resistance of *Salmonella* strains isolated from naturally polluted waters, *Journal of Applied Bacteriology*, **68**, 297–302.

Oliver, J.D., Hite, F., McDougald, P. *et al.* (1995) Entry into, and resuscitation from, the viable but non culturable state by *Vibrio vulnificus* in an estuarine environment, *Applied and Environmental Microbiology*, **61**, 2624–30.

Oremland, R.S. (1988) Biogeochemistry of methanogenic bacteria, *Biology of Anaerobic Microorganisms* (ed. A.J.B. Zehnder), pp. 641–705, John Wiley, New York.

Pedersen, H., Lomstein, B.A. and Blackburn, T.H. (1993) Evidence for bacterial urea production in marine sediments, *FEMS Microbiology Ecology*, **12**, 51–9.

Pfennig, N. (1988) Ecology of phototrophic purple and green sulphur bacteria, *Autotrophic Bacteria* (eds. H.G. Schlegel and B. Bowien), pp. 97–116, Springer-Verlag, New York.

Ploug, H., Lassen, C. and Jorgensen, B.B. (1993) Action spectra of microalgal photosynthesis and depth distribution of spectral solar irradiation in a coastal marine sediment of Lim Fjiorden,

Denmark, *FEMS Microbiology Ecology*, **12**, 69–78.

Rath, J. and Herndl, G.J. (1994) Characteristics and diversity of β-D-glucosidase (EC 3.2.1.21) activity in marine snow, *Applied and Environmental Microbiology*, **60**, 867–73.

Rattray, J. and Logan, N.A. (1993) Effects of pH and temperature on heterotrophic bacteria in acidified and non-acidified lochs, *Journal of Applied Bacteriology*, **75**, 283–91.

Schulzte-Lam, S. and Beveridge, T.J. (1994) Nucleation of celestite and strontianite on a cyanobacterial S-layer, *Applied and Environmental Microbiology*, **60**, 447–53.

Sinigalliano, C.D., Kuhn, D.N. and Jones, R.D. (1995) Amplification of the *amoA* gene from diverse species of ammonium-oxidizing bacteria and from an indigenous bacterial population from seawater, *Applied and Environmental Microbiology*, **61**, 2702–6.

Sommaruga, R. and Psenner, R. (1995) Permanent presence of grazing-resistant bacteria in a heterotrophic lake, *Applied and Environmental Microbiology*, **61**, 3457–9.

Stott, J.F.D. (1988) Assessment and control of microbially-induced corrosion, *Metals and Materials*, **4**, 224–9.

Sugita, H., Okamoto, N., Nakamura, T. *et al.* (1993) Characterization of microaerophilic bacteria isolated from the coastal waters of Tokyo Bay, Japan, *FEMS Microbiology Ecology*, **13**, 37–46.

Taylor, B.F. (1992) Marine habitats, bacteria, *Encyclopaedia of Microbiology*, Vol. 3 (ed. J. Lederberg), Academic Press, New York.

Tranvik, L.J. (1993) Microbial transformation of labile dissolved organic matter into a humic-like matter in seawater, *FEMS Microbiology Ecology*, **12**, 177–83.

Ulitzur, S., Reinhertz, A. and Hastings, J.W. (1981) Factors affecting the cellular expression of bacterial luciferase, *Archives in Microbiology*, **145**, 342–6.

Van Gemerden, H., Tughan, C.S., de Wit, R. and Herbert, R.A. (1989) Laminated mat microbial ecosystems on sheltered beaches in Scapa Flow, Orkney Islands, *FEMS Microbiology Ecology*, **62**, 87–102.

Varnam, A.H. and Evans, M.G. (1991) *Foodborne Pathogens: An Illustrated Text*, Wolfe Publishing Ltd, London.

Verran, J., Stott, J.F.D., Quarmby, S.L. and Bedwell, M. (1995) Detection, cultivation and maintenance of *Galionella* in laboratory microcosms, *Letters in Applied Microbiology*, **20**, 341–4.

Zehr, J., Mellon, M., Braun, S. *et al.* (1995) Diversity of heterotrophic nitrogen fixation genes in a marine cyanobacterial microbial mat, *Applied and Environmental Microbiology*, **61**, 2532–2537.

Zubkov, M.E. and Sleigh, M.A. (1995) Bacterivory by starved marine heterotrophic nanoflagellates of two species which feed differently, estimated by uptake of dual radioactive-labelled bacteria, *FEMS Microbiology Ecology*, **17**, 57–66.

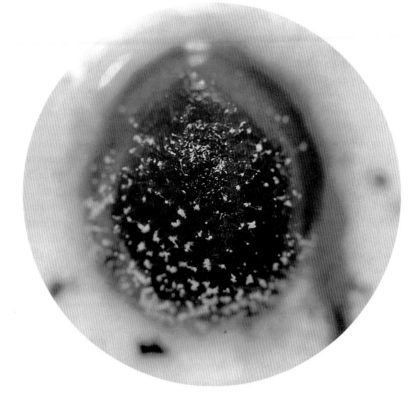

3 TERRESTRIAL ENVIRONMENTS

In the current context the terrestrial environment includes the land itself, with special reference to soil and habitats above and below the soil layer, including plants and buildings. This obviously encompasses a huge range of habitats, from fertile soils to barren rocks and including the highly productive agricultural regions as well as extremely arid deserts, both hot and cold. As with the aquatic environment, the success of micro-organisms in colonizing the most inhospitable niches is remarkable. These inevitably include the works of man. Equally, however, the cultivation of soils and other human activities means that man can impinge on micro-organisms to a greater extent than in aquatic environments and far greater effort is made to influence the microflora.

SOIL

Although there are many areas where the original rocks appear on the surface, a great part of the land surface is covered by rock particles in a superficial layer of varying thickness, which forms the skeleton of the soil. Soil is a complex structure and the range of living organisms, including micro-organisms, varies considerably. All soils possess a profile, often described in terms of 'horizons'. Horizon A is the upper layer and hence the most weathered, horizon B is less weathered but contains fine soil particles and soluble substances washed in from above, while horizon C is practically unweathered. Weathering is a continual process, important in providing a supply of neutralizing bases and stabilizing the soil pH value. Very hard rocks, such as granite, weather very slowly and the supply of bases is inadequate to prevent acidification.

Soil in horizon A, which is of the greatest importance in terms of agricultural activities, contains considerable quantities of organic matter as well as mineral constituents. The organic content of soil primarily consists of incompletely decomposed plant tissues as humus (see page 99). In addition, a greater or lesser amount of material of animal origin will be present, together with undecomposed material, the litter. Humus is in a steady state, being constantly replenished by the incorporation of organic residues, while other parts are oxidized to completion. As a consequence the humus content is highest in soils, such as cold prairies, where conditions are favourable for its formation and unfavourable for its decomposition. In contrast, humus is rapidly degraded in tropical soils, which usually have a correspondingly low content. The nature of humus also varies according to a number of factors. Extreme types are the mor humus of acid soils of low calcium content and the mull humus of higher calcium content and lower acidity.

Humus plays an important role in soil stability, mull humus acting as a natural high molecular weight ion exchange resin which maintains an ionic equilibrium. Humus-rich soil also maintains a large microbial population, which is itself of importance in enhancing soil stability.

The physical structure of soil is a network of pores which, unless waterlogged, permits gaseous exchange. The soil atmosphere is higher in CO_2 and lower in O_2 than the external atmosphere due to microbial respiration. Except when very dry, soil contains a film of moisture over soil particles and free moisture in the pores.

Many types of soil support a large and complex population of micro-organisms. It is recognized that conventional cultural methods are inadequate to detect and recover all types and alternative methods, such as fluorescent staining (156), as well as serological and genetic techniques are now in use. Bacteria and moulds (microfungi) are most important, the latter being favoured by low pH values. Yeasts are usually relatively unimportant except in specific situations such as vineyard or orchard soils. Protozoa are present as predators and there is a large bacteriophage population, although a significant proportion of this is probably present in the

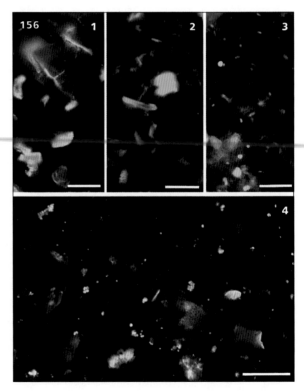

156 Use of a fluorescent microscopy technique to visualize living bacteria in soil. (1) Unstained fresh soil. (2) Fresh soil stained with sulfofluorescein acetate (SFDA).Microbial cells exhibit a yellow/green fluorescence. (3) SFDA added to autoclaved soil. (4) SFDA-stained soil after addition of vegetative cells of Bacillus subtilis. All bars equal 40 µm. Reproduced with permission from Tsuji *et al.*, 1995, © 1995, American Society for Microbiology.

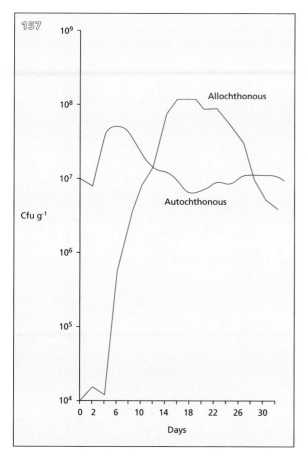

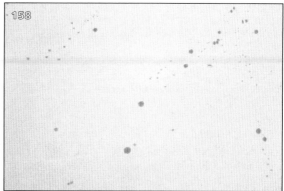

158 *Agromyces rasmosus*. This is a Gram-positive bacterium that consists of branched, filamentous elements, 1 µm or less in diameter. Dilute media are required for the isolation of *Ag. rasmosus* and for initial sub-culturing. (Source: heavily manured field soil; Medium: mineral salts plus soil extract; Incubation: 25°C (77°F), 12 days.)

157 Effect of an increase in nutrient content on the autochthonous and allochthonous microflora of soil. The events illustrated followed the application of a dilute cattle slurry to chalk soil c.4 weeks after harvesting malting barley. The weather was dry at the time of application (day 2), but light rain fell on days 4 and 6 and heavy, intermittent rain from days 10–14 and 18–20. Soil was sampled at a depth of c.3 cm and counts made on soil extract agar medium incubated at 22°C (71.6°F). Autochthonous and allochthonous microflora were differentiated on the basis of cell morphology and Gram reaction, pigmentation and simple biochemical characteristics.

temperate state, which is considered to be a mutualistic relationship (Marsh *et al.*, 1993). It is common practice to describe the microflora in soil in terms of two groups, the autochthonous and the allochthonous (zymogenous) micro-organisms. The autochthonous microflora of the soil is always present in significant numbers, irrespective of the nutrient status. In contrast, the allochthonous microflora is dependent on a periodic increase in nutrient status and, under normal conditions, is present only in very small numbers or present as either exospores or endospores. Steep nutrient gradients in soil mean that autochthonous and allochthonous micro-organisms coexist under some circumstances. Response to a local increase in nutrient level is rapid and colonization of minute pieces of organic material by allochthonous micro-organisms can occur immediately adjacent to much larger nutrient-poor areas. It should also be appreciated that members of the autochthonous microflora are likely to be in a low state of metabolic activity for extended periods. A number of microbiologists have referred to viable, non-culturable organisms in soil, although this term should, as ever, be used with caution. The autochthonous microflora is, however, able to respond rapidly to an increased availability of nutrients. Where increases are small the autochthonous microflora benefits but large increases in nutrient availability favour the faster growing allochthonous micro-organisms (**157**).

Members of the autochthonous microflora include both heterotrophic micro-organisms and specialist groups, such as ammonia-oxidizing bacteria. In some soils the most numerous micro-organism is *Agromyces rasmosus* (**158**) although the importance of this micro-organism is often underestimated.

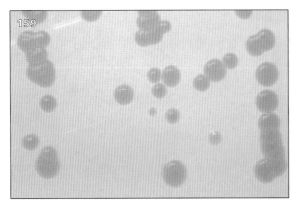

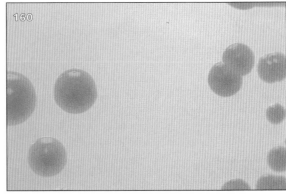

159 A species of *Arthrobacter*. *Arthrobacter* can easily be isolated from soil by direct plating on to a non-selective medium such as soil extract agar. (Source: pasture soil; Medium: soil extract agar; Incubation; 25°C (77°F), 3 days.)

160 A lemon-pigmented *Arthrobacter* isolate. Although most strains of *Arthrobacter* are buff or white in colour, yellow pigmentation is not unusual. (Source: aged chalk spoil heaps; Medium: soil extract agar; Incubation: 25°C (77°F), 5 days.)

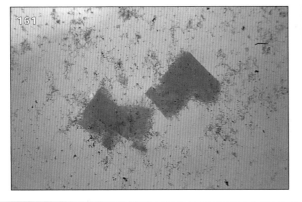

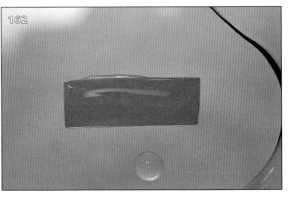

161, 162 Isolation of *Cellulomonas*. Although *Cellulomonas* can sometimes be isolated by direct plating on to a medium containing finely divided cellulose, enrichment is more reliable. An effective method is to inoculate a mineral salts medium containing a sterile strip of filter paper with a soil suspension. The presence of cellulolytic bacteria is indicated by the disintegration of the filter paper (**161**); control shows no disintegration (**162**). Positive enrichments are streaked on to cellulose agar; the developing colonies are examined for zones of clearing of the cellulose. (Source: woodland soil; Medium: mineral base with filter paper; Incubation; 28°C (82.4°F), 12 days.)

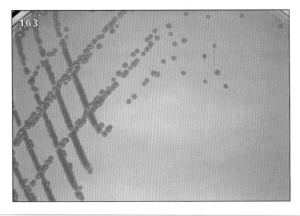

163 A species of *Cellulomonas*. The yellow pigmentation of *Cellulomonas*, together with its cellulolytic properties, provide useful clues in confirmation of its identity. Isolation procedures are not specific for *Cellulomonas* and further testing is required. (Source: as for **161**.)

Other Gram-positive bacteria, which may still be loosely referred to as 'coryneforms', are also of major importance in soil. These include *Arthrobacter* (**159**, **160**) and *Cellulomonas* (**161–165**). Gram-negative, rod-shaped bacteria are also important members of the autochthonous microflora. Oxidase-positive genera, such as *Pseudomonas* and *Janthinobacterium* (**166–169**), are often numerically dominant, but a wide range of other bacteria are present in large numbers. Members of the *Enterobacteriaceae* are often considered only in terms of their relationship with plants but some isolates, including *Erwinia herbicola* (**170**), also are considered to be part of the autochthonous microflora.

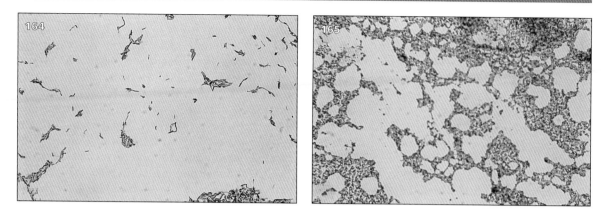

164, 165 Although *Cellulomonas* has a 'coryneform' morphology (**164**), the full cell cycle of genera such as *Arthrobacter* is usually absent. In the example illustrated, there is a trend to a coccoid morphology after 6 days' incubation (**165**) but many strains exhibit only limited shortening. (Source: as for **161**; Microscopy: Gram stain, × 1,200.)

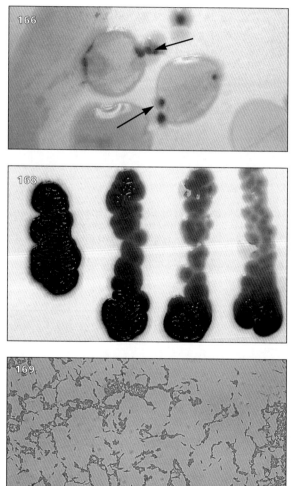

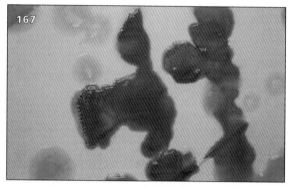

166–168 The three ages of *Janthinobacterium*. The violet pigment violacein is produced slowly by *Janthinobacterium* and initially appears as small dots (**166**; arrowed). A pale yellow fluorescent pigment may also be produced but this is not a universal characteristic. Pigment then spreads as a pellicle across the rubbery surface of the maturing colony (**167**). Senescent colonies are fully pigmented, the pigment darkens with further ageing and the colony surface becomes deeply wrinkled (**168**). (Source: soil, riverside park; Medium: nutrient agar; Incubation: 25°C (77°F), 2, 4 and 7 days.)

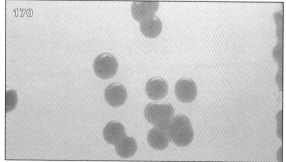

169 *Janthinobacterium lividum*. The cells are variable in size with rounded ends and may be slightly curved or irregular. Cells of significantly greater length are occasionally present in older cultures. (Source: as for **166**; Microscopy: Gram stain, × 1,200.)

170 A strain of *Erwinia herbicola*. *Erwinia herbicola* comprises various yellow and non-pigmented bacteria isolated from both healthy and diseased plants, man and animals and, less commonly, soil and water. (Source: field soil; Medium: nutrient agar; Incubation: 30°C (86°F), 2 days.)

As discussed above, the distinction between autochthonous and allochthonous micro-organisms is by no means absolute. Soil is a secondary habitat for *Micrococcus* (**171**) but the organism may be part of either the autochthonous or allochthonous microflora. Equally, closely related strains of the same organism can be members of either the autochthonous, or allochthonous microflora, depending on minor phenotypic differences.

It has been stated that hyphae-forming micro-organisms, fungi (**172**, **173**) and Actinomycetes (**174**, **175**) are the best adapted to benefit from increased nutrient concentrations in soil. This is true in that hyphae of both fungi and Actinomycetes are able to colonize discrete pieces of organic matter very rapidly and are able to form 'bridges' between nutrient-rich areas. In each case exospores (**176–182**) are of importance both as a means of distribution and persistence during periods of non-growth. *Streptomyces* is of particular note both for production of medically important antibiotics (such as chloramphenicol and the eponymous streptomycin) and for the production, along with the gliding bacterium *Nannocystis*, of volatile compounds responsible for

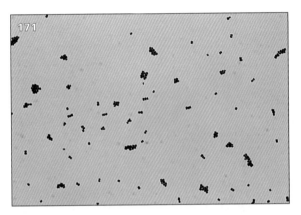

171 A species of *Micrococcus*. Although *Micrococcus* is primarily an inhabitant of the mammalian skin, the organism is relatively common in its secondary habitats of soil, water, milk and other foods. (Source: field soil; Microscopy: Gram stain × 1,200.)

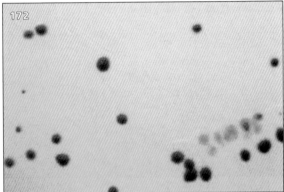

172 *Cladosporium herbarum*. Many moulds can be isolated from soil and often require suppression when examinations are being made for bacteria. *Cladosporium* is unusual in forming discrete colonies on solid media. This trait is reflected in its spoilage pattern on foods and growth as black spots on walls and other materials. (Source: dust in garden; Medium: malt extract agar; Incubation: 20°C (68°F), 5 days.)

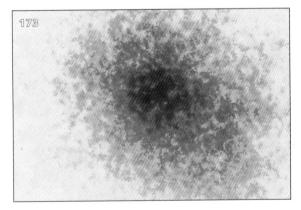

173 A single colony of *Cladosporium herbarum*. The dense nature of the olive-green mycelium is apparent in the centre of the colony. Spore heads consisting of tree-like clusters of conidia are present. (Source: as for **172**; Microscopy: unstained, 100 × bright field.)

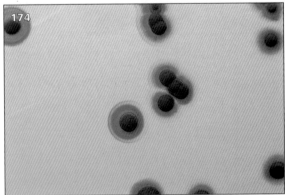

174 A species of *Streptomyces*. Although there is considerable variation in colony appearance, colonies of *Streptomyces* are easily recognized with a little experience. (Source: garden soil; Medium: starch–casein agar supplemented with actidione and nystatin; Incubation: 22°C (71.6°F), 5 days.)

175 Vegetative mycelium of *Streptomyces griseus*. The highly branched mycelium is very effective in penetrating and colonizing discrete pieces of organic matter. (Source: as for **174**; Microscopy: Gram stain, × 40.)

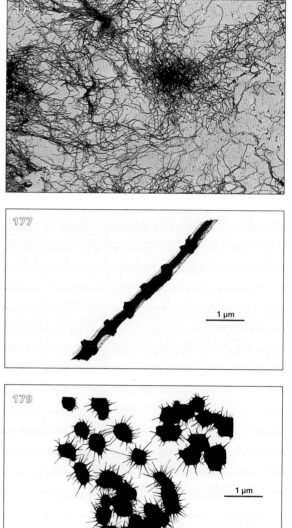

176–181 Morphology of the spores of six species of *Streptomyces*. Spores of *Stm. cacaoi* (**176**) are smooth but may easily be distinguished from those of *Stm. aureofaciens* (**177**) which are of a 'phlangiform' type. *Stm. hirsutus* (**178**) and *Stm. fasciculatis* (**179**) both produce spiny spores, the latter being appreciably longer. Spores of *Stm. griseoplanus* (**180**) are warty in appearance while those of *Stm. flavoviridis* (**181**) are hairy.

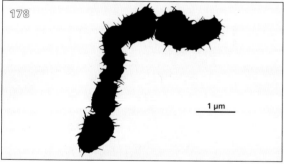

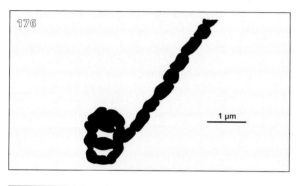

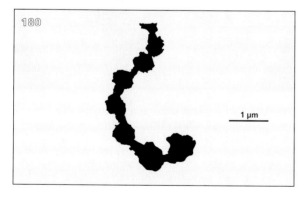

182 Spore chain of *Actinomadura madurae*. Soil isolates of *Actinomadura* develop aerial mycelia and superficially resemble *Streptomyces* on isolation plates. Examination by low power microscopy is required to differentiate between the organisms. Short or, occasionally, long chains of arthrospores are formed which may be straight, hooked, or irregular spirals.

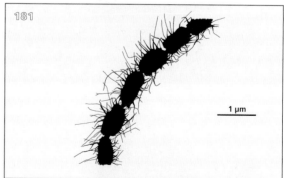

the smell of soil (**183**, **184**). Systematic information concerning Actinomycetes is surprisingly difficult to obtain although a study in China's Yunnan province (Xu *et al.*, 1996) showed that *Streptomyces* represented 90% of the population of Actinomycetes; *Streptomyces* was most prevalent in primeval forest, followed by secondary forest and vegetative farmland.

Gliding bacteria may be present in soil in large numbers. These include fruiting myxobacteria (**185–195**), the most numerous of which is usually *Nannocystis*, and non-fruiting myxobacteria, including

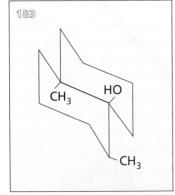

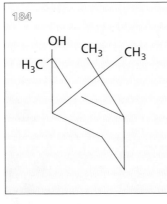

183, **184** Structure of volatile compounds produced by *Streptomyces*. A number of volatile compounds have been associated with the smell of damp earth, of which geosmin (**183**) and 2-methylisoborneol (**184**) are considered of greatest importance.

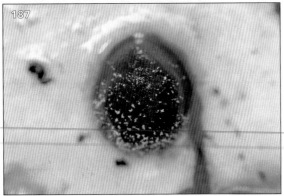

185–187 Two methods of baiting for bacteriolytic myxobacteria. In the first (**185**), soil is moistened to a muddy consistency with actidione (cycloheximide) solution. A number of autoclaved, antibiotic-free rabbit pellets are pressed into the soil and the plate incubated at room temperature. The second method (**186**) involves soaking a pad of two or three filter papers with actidione solution. The material under examination, bark, decaying vegetable material or, in this case, rabbit pellets is placed on the filter paper and the plate incubated at room temperature. Myxobacteria may be recognized as small, white, or brightly coloured fruiting bodies (**187**).

188–191 Fruiting body formation in *Chondromyces*. The aggregation of cells (**188**) is followed by the formation of a fruiting body (**189, 190**). The mature fruiting body (**191**) consists of a mucous stem and cysts. Cysts are organs of distribution and release myxospores on germination. These develop into new vegetative cells. (All figures approximately x 10–x 15.)

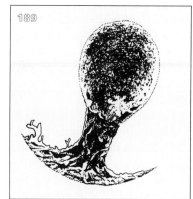

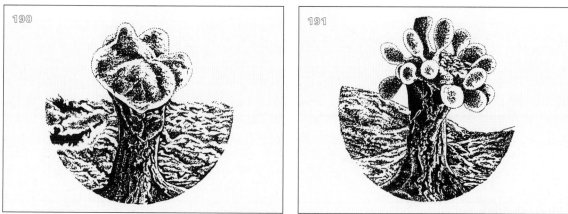

192–195 Fruiting bodies of various myxobacteria. *Melittangium lichenicolum* (**192**); *Stigmatella aurantiaca* (**193**); *Cystobacter fuscus* (**194**); *Polyangium* sp. (**195**).

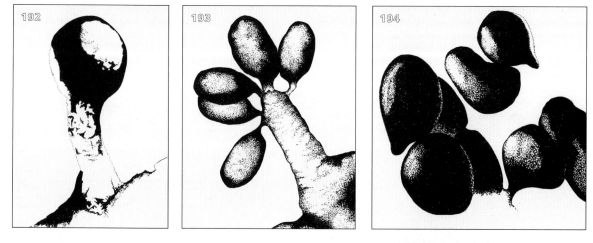

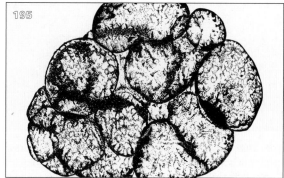

Lysobacter and members of the *Cytophaga* group (196–199). Gliding strains of bacteria not normally associated with this trait are not uncommon and gliding may well be a general property of soil bacteria rather than a specific property of a limited number of genera.

Endospore-forming bacteria are very common in soil and often dominate isolation plates. *Bacillus* (200–202) is the most common genus and an important member of the allochthonous microflora, although its importance can easily be overestimated as a consequence of its persistence in soils. The animal and human pathogen *B. anthracis* is common in soils in some areas of Asia and Africa while the closely related *B. cereus*, which is a cause of food poisoning and extra-intestinal disease, is very widespread. *Sporosarcina ureae* can be isolated in large numbers from some alkaline soils and from soils where dogs urinate (203). Urea hydrolysis by *Sp. ureae* is of economic importance where urea is used as a base for fertilizers and soil inoculation with the organism has been practised on a commercial basis.

Oxygen gradients within the soil structure are such that aerobic, micro-aerophilic and anaerobic

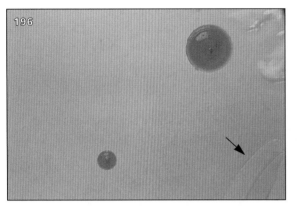

196 Isolation of *Cytophaga*. *Cytophaga* is present in large numbers in many environments and may be isolated by direct plating. In general, a medium of low nutrient content is used to encourage swarming and reduce competition. The swarm is most easily observed in the area where a loopful has been removed for microscopic examination (arrowed). In addition to swarming, *Cytophaga* is recognized by yellow-red pigmentation and cell morphology, although none of these characters is fully dependable. (Source: uncultivated soil; Medium: *Cytophaga* agar; Incubation: 20°C (68°F), 5 days.)

197 *Cytophaga* species are of differing cellular morphology, varying from long slender rods to short, almost coccal rods. There is some indication of a cell cycle in many types of swarming bacteria and in the illustration some of the long rods are breaking down. (Source: as for **196**; Microscopy: Gram stain, × 1,200.)

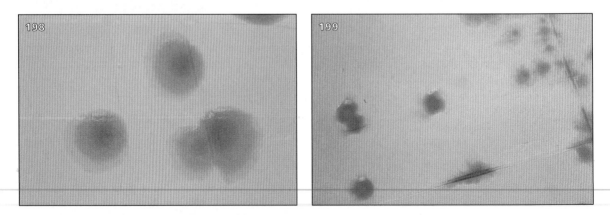

198, 199 The ability of *Cytophaga* to swarm on laboratory media is dependent on peptone concentration. The organism is able to swarm on a medium containing 0.075% peptone (**198**) but not at a peptone concentration of 0.15% (**199**). A concentration of 0.1% is generally considered critical.

growth can occur in very close proximity. In most soils *Clostridium* is the most important anaerobic genus although, as with *Bacillus*, persistence of endospores means that the importance of members of this genus can be overestimated. The dangerous pathogen *Cl. tetani* (**204**) is common in many soils and the ease of isolation underlines the importance

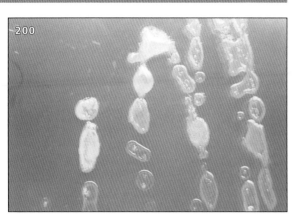

200 Young colonies of *Bacillus licheniformis*. Growth may be visible within 12 hours under suitable conditions. (Source: garden soil; Medium: nutrient agar; Incubation: 37°C (98.6°F), 15 hours.)

201 *Bacillus subtilis*. *Bacillus subtilis* is common in dried foods such as flour, as well as in soil. The organism is responsible for spoilage of bread (rope) and also a cause of mild food poisoning. The decreased use of antimicrobial agents in bread has led to an increase in rope spoilage and also cases of food poisoning. (Source: grassland soil, Medium: blood agar; Incubation: 37°C (98.6°F, 24 hours.)

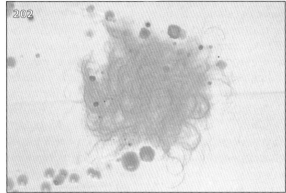

202 *Bacillus mycoides*. *Bacillus mycoides* may be identified on the basis of the characteristic colonial morphology with a high degree of probability. (Source: garden soil; Medium: nutrient agar; Incubation; 30°C (86°F), 2 days.)

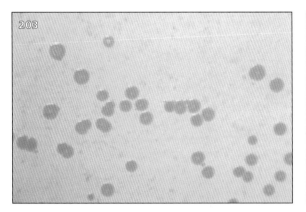

203 *Sporosarcina ureae*. Isolation is made on nutrient agar containing at least 3% urea to suppress *Bacillus* species. Colonies are initially white or grey but may become yellowish, orange or brown with ageing. (Source: soil, where dogs urinate; Medium: nutrient agar containing 3% urea; Incubation: 25°C (77°F), 2 days.)

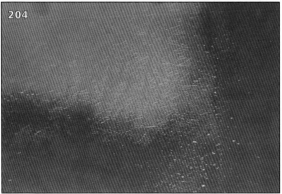

204 *Clostridium tetani* often produces spreading colonies that consist of a flat network with a delicate edge of projecting filaments. Haemolysis may be seen in the older dense part of the colony. (Source: arable soil; Medium: blood agar; Incubation: 37°C (98.6°F), 2 days, 95% H_2:05% CO_2.)

of immunization against the organism for children and other persons at risk, including agricultural workers. The pathogen *Cl. perfringens* type A (**205, 206**) is also common in soil, although the primary habitat is the animal intestine. Other groups of *Cl. perfringens*, however, are obligate parasites and so are rare in soil.

Heterotrophic bacteria in soil are of very considerable importance in mineralization and recycling of matter. A considerable amount of the organic input into soil is plant material, which contains high quantities of large complex molecules. Cellulose, a basic component of all plant materials, comprises 40–70% of plant residues in soil and the importance of cellulolytic micro-organisms in mineralization and the carbon cycle is obvious. Fungi are of greatest importance in acid soils and are also able to degrade cellulose in wood, where it is embedded in lignin. Species of *Fusarium* and *Chaetomium* are predominant but several other genera, including *Aspergillus*, *Botrytis* and *Rhizoctonia*, are also cellulolytic. Amongst bacteria, *Cytophaga* and *Sporocytophaga* are the most important cellulolytic genera in aerobic soils and are often predominant in soils of neutral or alkaline pH. The extent of attack is truly remarkable and is characterized by

the necessity for direct contact between the bacterium and the cellulose substrate. It is likely that cellulolytic activity is primarily due to a cell-bound exoenzyme.

A variety of other bacteria possess cellulolytic activity but the importance in soils is difficult to assess. Fruiting myxobacteria of the genera *Polyangium*, *Sorangium* and *Archangium* have been demonstrated to be cellulolytic and the property is also present in a few species of *Streptomyces* and in *Micromonospora chalcea*. *Cellulomonas* is well known, although of considerable less importance in soil than cytophagas. A process has been developed for the use of *Cellulomonas* in the production of protein from waste cellulose but does not appear to have been adopted commercially. Cellulolytic activity has more recently been reported in *Pseudomonas* and it is possible that the property is more widely distributed in terrestrial bacteria than thought previously. Cellulase production, however, is often subject to catabolite repression and the enzyme is not usually produced where other substrates are plentiful.

Both mesophilic and thermophilic *Clostridium* species are involved in cellulose hydrolysis in anaerobic soils and the process is of major importance

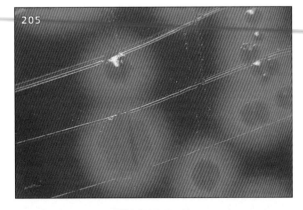

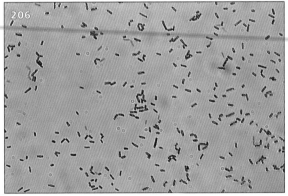

205 Colonies of *Clostridium perfringens* type A often have a target appearance on blood agar. The inner haemolytic zone is due to theta toxin and the outer zone of partial haemolysis to alpha toxin. Strains of *Cl. perfringens* type A, associated with food poisoning, are often non-haemolytic and colonies lack the inner ring of complete hydrolysis. A zone of partial clearing is usually present due to lecithinase activity. Lecithinase-negative strains of *Cl. perfringens* have been implicated in outbreaks of food poisoning. Colonies of these strains on blood agar have no distinguishing features and for this reason may be overlooked. (Source: garden soil; Medium: blood agar; Incubation: 37°C (98.6°F), 2 days, 95% H_2:5% CO_2.)

206 *Clostridium perfringens*. Experienced persons can recognize this organism on the basis of cellular morphology. The cells are relatively short, stout and square-ended. Endospores are not usually seen. In some cases cells are arranged in lines in 'box-car' formation. Those illustrated are rather less regular than usual. (Source: as for **205**; Microscopy: Gram stain × 1,200.)

in landfill. Products of the cellulose fermentation, including acetate and H_2, are important substrates for methanogens and sulphate-reducing bacteria. Recent thinking suggests that acetate may form a trophic link between anaerobic and aerobic processes (Kusel and Drake, 1995).

Many other plant materials, including xylan, starch and glucans, are more readily metabolized than cellulose. The obvious exception is lignin, which accounts for 18–30% (dry weight) of woody material. Lignin is the most recalcitrant plant material and is a major source of humus. Lignin is metabolized by some fungi, all of which are members of the Basidiomycetes. These include *Polystictus versicolor* and *Stereum hirsutum*, which preferentially metabolize lignin, and *Amillaria mellea*, *Pleurotus ostreatus* and *Polyporus adustus* which metabolize lignin simultaneously with cellulose. Fungi of this type are often associated with rotting of fallen wood but some can degrade lignin in the living plant. *Amillaria mellea*, for example, is an important pathogen of both coffee and tea plants. Lignin can also be degraded by some Gram-negative bacteria including *Agrobacterium*, *Flavobacterium* and *Pseudomonas* and possibly by *Streptomyces*. In all cases, however, metabolism is very slow.

Humus is the name commonly applied to amorphous organic compounds in soil; it also contains recalcitrant materials. Lignin is the major component but fats, waxes, carbohydrates and proteins are also present. These are converted into poorly defined polymeric substances by processes involving bacteria, fungi, protozoa, nematodes and worms (**207**). A key feature of the formation of humus is the accumulation of nitrogen, the C:N ratio of c.40:1 in plant tissue falling to c.10:1 in humus.

Higher animals are also a source of substrates for heterotrophic activity in soil, either through production and dispersal of faeces and urine or through death. Chitin, primarily derived from the cell walls of Basidiomycetes and Ascomycetes, is also an important substrate, soil populations of up to 10^6 chitinase-producing micro-organisms suggesting a constant supply. Many genera of bacteria and fungi produce chitinase, including *Bacillus*, *Cytophaga*, *Nocardia*, *Micromonospora*, *Pseudomonas*, *Streptomyces*, *Aspergillus* and *Mortierella*. Fruiting bodies of Basidiomycetes may themselves be colonized before death (**208**).

The wide heterotrophic activity of soil microorganisms is of very considerable importance in the remediation of soil after pollution with hydrocarbons and other organic chemicals. Long-chain hydrocarbons (alkanes) can be metabolized by a wide range of bacteria, fungi and yeasts, although bacteria are of greatest importance in soils. The chain length is of major importance in deciding the number of species able to metabolize a given hydrocarbon.

Species of *Mycobacteria* and *Nocardia* are able to metabolize longer chain hydrocarbon molecules although recalcitrant residues may remain. *Pseudomonas* and other Gram-negative aerobes are involved in, e.g. the oxidation of petroleum.

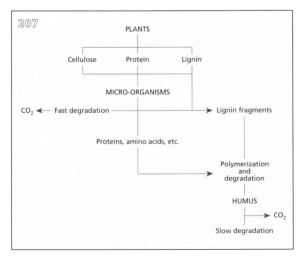

207 Stages in the formation of humus.

208 Growth of *Pseudomonas* on fruiting bodies of a Basidiomycete. Discrete colonies are present, but *Pseudomonas* was present in large numbers in a slime matrix on the surface of the fruiting body and its immediate surroundings.

Degradation is often complete in well-aerated soils (*Feature 8*).

The apparent ability of micro-organisms to degrade all organic carbon compounds led to the principle of 'microbial infallibility' which indicated that, sooner or later, all molecules would be degraded. There are, however, exceptions amongst synthetic compounds, some of which appear to be almost totally resistant to degradation by micro-organisms.

Amongst the most important of these are pesticides, DDT persisting for about 10 years and chlordane for over 12 years in aerobic environments. Degradation in anaerobic environments is likely to be significantly slower. The widespread use of synthetic pesticides leads to contamination of the general environment including, inevitably, aquifers supplying drinking water.

The environmental damage caused by the use of synthetic materials highly resistant to degradation is now widely recognized. Use of the most persistent agricultural chemicals is avoided if possible and biodegradable plastics have replaced their recalcitrant predecessors in many situations. Long-term environmental damage not only remains as a legacy of less enlightened days but still occurs as a result of accident, carelessness and greed. The effect of pollution on soil microflora (**209**) is varied but in a number of cases it has been possible to exploit the metabolic versatility of micro-organisms in remediation.

Soil micro-organisms play a very important role in the nitrogen cycle. Nitrogen fixation is generally considered to be the rate limiting step in nitrogen turnover and is a property unique to the realm of prokaryotes. Total global nitrogen turnover has been estimated at 10^8–10^9 tons per year. It has been calculated that micro-organisms are responsible for significantly more than 50% of all nitrogen fixed. The only other major source is the Haber–Bosch process used for the industrial production of fertilizers, other non-biological sources, such as

---

Feature 8. Desert storm – the aftermath

Reporting of the Gulf War largely concentrated on military matters and the related geopolitical consequences. A lesser known consequence was the massive scale of oil pollution, both of the coastline and of inland desert soil. The onset of bioremediation on the coastline was marked by the development of cyanobacterial mats on the surface of the oil. Filamentous cyanobacteria were dominant – *Phormidium* spp. in Kuwait and *Microcoleus* spp. in Saudi Arabia. The mats served to immobilize oil-degrading bacteria of which *Rhodococcus* spp. was dominant, but *Bacillus* and *Arthrobacter* were also involved.

A bioremediation programme for contaminated soil involved the use of pure cultures of the oil-degrading bacteria isolated from the cyanobacterial mats. This microflora remained dominant for 4 weeks but, between 4–18 weeks, species of *Pseudomonas* increased in number to share the dominance. *Streptomyces* and a bacterium 'resembling' *Thermoactinomyces*, though not thermophilic, also increased in number to significant levels. Some degree of colonization by the moulds *Aspergillus* and *Penicillium* also occurred. Both streptomycetes and moulds, however, declined in numbers towards the end of the degradation cycle. The microbial consortium was able to degrade 50% of the oil over a 10–20 week period. Data from Sarkoh *et al.*, 1995.

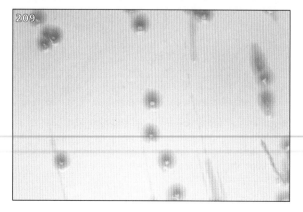

209 Effect of pollution on soil microflora. Spillage of an agricultural chemical (nature not known) on a small area of chalk soil resulted in a rapid fall in numbers (determined by plating on soil extract agar) from c.10^8 cfu g^{-1} to c.10^7 cfu g^{-1}. Numbers increased to pre-spillage levels over a 2–3 week period, during which rainfall was minimal. There was no observed effect on diversity or the relative proportions of different types of micro-organism, with the exception of a 'coryneform' bacterium of distinctive colonial morphology. This organism first appeared 2 weeks after spillage and has persisted for over 1 year at low and variable levels. (Medium: soil extract agar; Incubation: 25°C (77°F), 5 days.)

lightning strikes, electrical equipment and the internal combustion engine accounting for no more than about 0.5% of total fixation.

A wide range of bacteria are able to fix nitrogen; work with cyanobacterial marine mats suggests that the ability is more widely distributed amongst heterotrophic bacteria than considered previously. In the western hemisphere the most important source of fixation is *Rhizobium* and the closely related *Bradyrhizobium* growing symbiotically with leguminous plants. The legume symbiosis is also quantitatively the most efficient means of biological nitrogen fixation. Significant quantities of nitrogen are also fixed by the symbiotic relationship between the actinomycete *Frankia* and non-leguminous plants. The contribution from nonsymbiotic bacteria is considered small on an overall basis but can be of considerable localized importance. The most important genus is *Azotobacter* although the closely related *Beijerinckia* is dominant in acid soils in the tropics. A considerable amount of N_2 is fixed in anaerobic soils by *Cl. pasteurianum* while *Arcobacter nitrofigilis* may be of importance where conditions favour micro-aerophilic life. In the eastern hemisphere, cyanobacteria (including *Anabaena azollae* in symbiotic relationship with the water fern *Azolla* are major sources of fixed N_2 while lichens (containing cyanobacteria as photo-bionts) are of major importance in cold deserts.

Nitrogen is fixed as ammonia, which can be used directly by the micro-organism, or its symbiont. Excess is released into the wider soil environment where, together with NH_4^+ produced from animal and plant sources, it is bound by clay particles and so largely unavailable. The process of nitrification involves the oxidation of NH_4^+ to NO_2^- and, subsequently, NO_3^-. As NO_3^-, nitrogen is freely diffusible and can be assimilated by plants and many soil bacteria. Microbiological oxidation of NH_4^+ involves specialist chemoautotrophic bacteria, the classic prototypes being *Nitrosomonas* (which oxidizes NH_4^+ to NO_2^-) and *Nitrobacter* (which oxidizes NO_2^- to NO_3^-). More recently, other bacteria which are physiologically similar to, but morphologically distinct from, the prototypes have been characterized. These include the NH_4^+ oxidizers *Nitrosococcus*, *Nitrosolobus* and *Nitrosospira* and the NO_2 oxidizers *Nitrococcus* and *Nitrospira*.

In the microbial world there are, of course, no 'free rides' – the release of N_2 is an essential part of the nitrogen cycle. Considerable amounts of nitrogen are stored as proteins in the microbial biomass and degradation to low molecular weight amino acids is important in the mineralization of nitrogenous compounds. Under most conditions, soil proteases are the rate limiting step in mineralization. It seems likely that soil proteases can be obtained from a wide range of sources. Plant debris is an important source in some circumstances, while actinomycetes are thought to be of prime importance in some soil types. In paddy soils (see page 104 for general discussion) species of *Bacillus* appear to be the most important source of proteases and this may also be true in other soils. Nitrogen is also present as proteins and nucleic acids in plants and animals and released into soil by mineralization after death. Animals also excrete nitrogenous compounds, the nature of which varies. Invertebrates primarily excrete NH_4^+, reptiles and birds uric acid and mammals urea. Uric acid and urea are both mineralized with the formation of NH_4^+ and CO_2. At the same time nitrate is lost by denitrification, a process which depletes the soil of plant nutrients and which is consequently detrimental to agricultural productivity. A wide range of bacteria may be involved and loss of nitrate can be rapid under conditions of reduced oxygen availability (*Feature 9*).

Denitrification is an essential part of the nitrogen cycle and the only process that converts fixed nitrogen to N_2. Nitrate is the end product of mineralization under aerobic conditions but the ion

---

Feature 9. All I need is the air that I breathe

Turnover of nitrogen pools, nitrite and ammonium can be very rapid, typically 24 hours in many grassland soils. Leaching of nitrogenous compounds from soils is a major problem, leading to water pollution, nutrient deficiencies for plants and the increased mobilization of aluminium, in itself a serious problem. Nitrate respiration, involving the reduction of NO_3 to NO_2 and, finally, dinitrogen gas, is usually an anaerobic process, nitrate or nitrite acting as the terminal electron acceptor in the absence of oxygen. Some bacteria, notably the Gram-negative *Escherichia coli* and *Pseudomonas aeruginosa* as well as the Gram-positive *Arthrobacter* are, however, capable of nitrate respiration under fully aerobic conditions. The significance of, and advantage to, the bacterium is not clear but co-respiration of nitrate and oxygen may make a significant contribution to the flux of nitrate to nitrite in the terrestrial environment. Data from Carter *et al.*, 1995.

is highly soluble and continuously leaches from soils into rivers and, ultimately, into the ocean.

In the absence of denitrification the entire nitrogen supply of the earth would accumulate in the oceans and terrestrial life, except on the ocean fringes, would cease.

Soil and climatic change

In many regions soil is subject to periodic dehydration for prolonged periods. Depending on other factors, many micro-organisms are relatively resistant to dehydration, although sub-lethal lesions may result, affecting the ultimate ability to recover. Coccoid micro-organisms are significantly more resistant than rod-shaped, both to dehydration *per se* and to effects such as increased salt concen-

trations. Survival is generally less at the higher ambient temperatures associated with dehydration. Under some conditions surface temperatures are sufficiently high for prolonged exposure to result directly in the death of some of the more sensitive micro-organisms. Bacterial endospores are highly resistant to both dehydration and high temperatures and, for this reason, endospore-forming bacteria are extremely common in soils. Only a relatively small number of genera form endospores but these spores appear to be important in permitting long-term environmental survival.

Exospores produced by microfungi and actino-mycetes are generally considered to be a means of dispersal rather than of survival. There is strong evidence, however, that exospores are of significantly greater resistance to desiccation than vegetative mycelia or the cells of non-mycelial micro-organisms. It is also possible that exospores are able to respond faster to an improvement in conditions than surviving but stressed bacterial cells.

Many species of the myxobacteria produce fruiting bodies containing spherical myxospores. Only a few genera have been studied in detail (primarily *Myxococcus*) but it has been found that myxospores are significantly more resistant to dehydration and UV irradiation (but only slightly more heat resistant) than vegetative cells and serve both as a resting structure and a means of dispersal.

Biofilms (see pages 8–12) would appear to be of considerable importance in protecting vegetative bacteria from dehydration. Production of extra-cellular polymeric substances, the major component of biofilms, is stimulated by desiccation (Roberson and Firestone, 1992). In this context, the biofilm can be regarded as a water-laden gel with a high capacity for moisture retention in adverse conditions. Under conditions of prolonged soil dehydration, water loss from the outer layers is inevitable with an associated loss of viability of some of the micro-organisms present. Desiccation does not destroy the inherent water-binding properties of the biofilm polymers and it is likely that the dried biofilm acts as a sponge, rapidly absorbing any moisture that becomes available (*Feature 10*).

Waterlogging is a common condition in some soils leading to the rapid development of anaerobic conditions. Depletion of oxygen leads to the soil taking on the metabolic characteristics of aquatic sediments, including methanogenesis and, where a sulphur source is present, sulphate reduction. Clostridia are responsible for the breakdown of cellulose and other large molecules. Cellulolytic clostridia include nitrogen-fixing species and, in the

---

Feature 10. After the fire a still small voice

Wildfires are increasing in number and becoming more widespread, possibly as a result of climatic change. The effect can be particularly dramatic on soil microflora as well as surface vegetation and animal life. Wildfires burn at very high temperatures and effect partial or total sterilization of the soil. Conditions immediately after a wildfire tend to be unfavourable to micro-organisms but after a short delay recolonization is rapid. A sharp increase to numbers significantly greater than those before the fire occurs as a result of nutrient release and favourable conditions of moisture and temperature. Heterotrophic bacteria gain the greatest advantage at the expense of autotrophic cyanobacteria and algae. Microbial numbers fall to pre-fire levels over a period of about one year although numbers of endospores remain high. The presence of a layer of fine ash and fire-induced soil compaction tends to create anaerobic conditions, the fall in numbers of aerobic bacteria being particularly marked. Extensive colonization by cyanobacteria and algae occurs where the removal of vegetation leads to a greater availability of light.

Microfungi are badly affected by wildfires with reduced propagule numbers and shortened hyphal length. Post-fire recovery of microfungi is relatively slow. From Vazquez *et al.*, 1993.

absence of leguminous plants, nitrogen fixation in water-logged soils can be significant. In contrast to most anaerobic environments, many waterlogged soils contain significant quantities of nitrate and denitrification is extensive though variable. Variation appears to be due primarily to differing rates of diffusion of substrates to the anaerobic centres of soil particles (Ambus, 1993). Streamside soils have a high capacity to support denitrification and may be important sinks for agricultural discharge. The extent and duration of waterlogging and the nature of the soil can make the consequences unpredictable (**210–214**).

210 A flooded part of a water meadow adjacent to a large inland river. An atypical *Bacillus* was present in the damp soil in large numbers but was not isolated from adjacent dry soil or the river water.

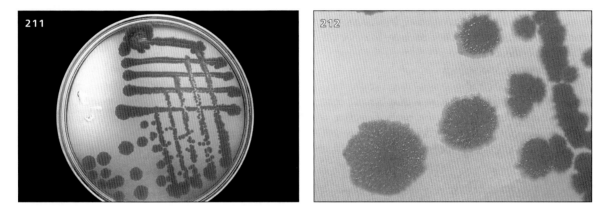

211, 212 An unidentified species of *Bacillus* producing a water-soluble brown pigment. This isolate could not be identified with any described species although phenotypically it appeared to be most closely related to *B. subtilis*. Colonial morphology is not entirely typical of *Bacillus* and cultures had an unusual 'sweet-floral' odour. (Source: soil, water meadow; Medium: nutrient agar; Incubation: room temperature, 4 days.)

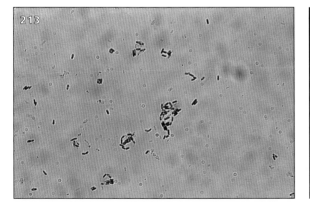

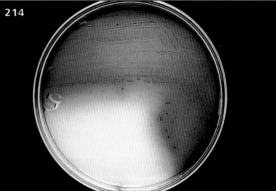

213 Unidentified brown pigmented *Bacillus*. Microscopic appearance is one of the phenotypic characteristics resembling *B. subtilis*. Cells are *c*.0.7 × 2.5 μm and endospores either do not swell the sporangium or produce a very slight swelling. (Source: as for **212**; Microscopy: Gram stain, × 1,200.)

214 *Listeria monocytogenes*. Although originally developed for use with clinical material, Oxford medium is effective for the isolation of listerias from foods and environmental specimens. Oxford agar recovers several species of *Listeria* in addition to *L. monocytogenes* and biochemical testing is required for confirmation. (Source: damp, riverside grassland soil. Medium: Oxford agar; Incubation: 37°C (98.6°F), 24 hours.)

Rice fields (paddies) are deliberately flooded and rapidly become anoxic. Methanogenic bacteria become active under anoxic conditions and so rice paddies are a significant source of atmospheric methane. During non-flooded periods, methanogens must survive desiccation in oxic conditions. Methanogens do not form any recognized resistant forms and it appears that pyrites (FeS_2) is an important factor in the protection of methanogens. The mechanism is not known but it is possible that the 'raspberry-like' surface (pyrites framboits) provides a protective habitat (Fetzer *et al*, 1993). The surface of pyrites appears to be of particular biological importance – recently it has been theorized that pyrites was the 'cradle of life', constituting the surface on which the first biomolecules were formed and around which the first cells developed (Wachtershauler, 1992). A similar protective mechanism may be operative in other soils subject to periodic desiccation and waterlogging.

The anaerobic metabolism of sulphur-containing compounds in rice paddies results in substantial H_2S production. Sulphide is toxic to a number of plants, rice seedlings being among the more sensitive. A symbiotic association has been described between rice plants and the sulphur-oxidizing bacterium *Beggiatoa*. *Beggiatoa* develops in the rhizosphere and significantly reduces H_2S concentrations, minimizing inhibitory effects on the rice. At the same time, oxygen, transported to the plant roots, diffuses into the rhizosphere creating micro-aerophilic conditions that favour *Beggiatoa*. *Beggiatoa*, which is catalase negative, also benefits from catalase production by plant roots.

Rice paddies are significant sites of biological nitrogen fixation. Cyanobacteria, primarily *Nostoc*, *Anabaena*, *Calothrix* and *Gloeotrichia* are of greatest importance and can fix as much as 40 kg of nitrogen per year. It is also notable that a high percentage of methanogenic bacteria in paddies are also capable of nitrogen fixation. There may also be a contribution from nitrogen-fixing species of clostridia. It has been suggested that the free-living, micro-aerophilic, nitrogen-fixing *Arcobacter nitrofigilis* may be of importance in rice fields and other flooded soils but evidence appears lacking. In any case the contribution made by eubacteria appears relatively small in comparison with that of cyanobacteria. Shortage of chemical fertilizers in developing countries means a considerable degree of interest in using cyanobacteria, such as *Calothrix*, as biofertilizer.

MICRO-ORGANISMS AND HIGHER PLANTS

Micro-organisms form a great many antagonistic, mutualistic and commensalistic relationships with plants. These are not always easily defined and some apparently commensalistic relationships are actually mutualistic. Relationships between micro-organisms and plants are of considerable economic importance, directly affecting agricultural productivity both through disease and increase in yields. Over the years considerable effort has been expended in manipulating the relationship between micro-organisms and plants in attempts to increase productivity. Concern over the environmental impact of agricultural chemicals such as pesticides has led to enhanced efforts to develop biological control agents. In some cases efforts are being hampered by the lack of a full understanding of the interactions between micro-organisms, plants and environmental factors.

The rhizosphere

The rhizosphere is the region of soil immediately surrounding the root hairs of higher plants together with the surface of the roots themselves. In the rhizosphere region, which effectively extends a few millimetres from the surface of each root, micro-organisms are directly influenced by the biochemical activities of the plant. Stimulation of the growth of micro-organisms leads to viable counts ten to several hundred times greater in the rhizosphere than in the surrounding soil. There is also an influence on the composition of the microflora, Gram-negative, rod-shaped bacteria appearing to obtain the greatest benefit. There are, however, no micro-organisms specific to the rhizosphere, all are also members of the microflora of non-rhizosphere soil.

Stimulation of bacteria is due to the release from plant roots of a vast range of organic materials, particularly carbohydrates, vitamins, amino acids and enzymes. Direct benefit to the plant is less easy to demonstrate but activities essential to plant growth, including mineralization and nitrogen fixation by free-living bacteria, are concentrated in the rhizosphere. The nitrogen-fixing *Azospirillum* forms associations with roots of cereals and other plants and has a direct beneficial impact on host growth, physiology and nutrition. The use of *Azo-spirillum* as biofertilizer is highly beneficial in situations where chemical fertilization is impractical, undesirable, or impossible (Pacovsky, 1990). Nitrite oxidation, however, can either be stimulated or suppressed by the presence of plant roots and it is possible that similar mechanisms control other activities. Other relationships are not straightforward and it is now appreciated that some saprophytic rhizosphere bacteria, primarily *Pseudomonas*, are harmful to plants. These 'deleterious rhizobacteria' (DRB) are not considered parasitic but induce deleterious effects on growth through the production of metabolites, candidate substances being hydrogen cyanide, unidentified toxins and phytohormones such as indole 3-acetic acid (Astrom *et al.*, 1993). The presence of DRB in soil has serious economic implications for crop production, especially in continuous cropping systems, unbalanced crop rotation and where seed is drilled directly into residue-rich soil. At the same time there is considerable interest in the use of DRB as a biological control agent for some weeds.

There is considerable competition amongst micro-organisms in the rhizosphere. This is of benefit to plants in that potential pathogens are often unable to establish themselves at the root surface. There is also evidence that some rhizosphere micro-organisms have a protective role against infestation by nematodes. There is considerable commercial interest in the use of rhizosphere micro-organisms, especially fluorescent pseudomonads, as biological control agents against root pathogens. There remains, however, a lack of understanding of the mechanisms by which these bacteria gain selective advantage. It seems likely that bacteria may employ various strategies including rhizosphere colonization and niche exclusion, ammonia production, production of secondary metabolites (such as antibiotics) and siderophore production, especially in alkaline soils where iron is a limiting element. Copper resistance is thought to be an important factor in the suppression of *Phytophthora parasitica* by pseudomonads in citrus grove soils, although both colonization of the fungal hyphae and siderophore production also appear to be necessary (Yang *et al.*, 1994).

Some normal inhabitants of the rhizosphere are also plant pathogens. These include some fungal pathogens and the bacterial pathogens *Agrobacterium tumefaciens* (**215**, **216**), the causative organism of crown gall disease (**217**) and *Ag. rhizogenes*, the causative organism of hairy root disease. *Agrobacterium* resembles *Rhizobium* (see page 110) in terms of morphology and physiology and there are similarities in the method of infection by the two micro-organisms (**218**). In many ways a comparison of *Agrobacterium* and *Rhizobium* illustrates the narrow division between parasitism and mutualism. *Agrobacterium* is distinguished from other plant pathogens in that crown gall tumours are capable of growing continuously in the absence of the initiator. This is due to a stable change in cell heredity, genes from the bacterial plasmids being transferred to, and integrated into, the host plant cells. There has been considerable success in exploiting this property and using *Agrobacterium* as vector in the production of transgenic plants (Evans, 1993).

The role of antibiotics in natural environments is always a matter of debate, some microbiologists considering production to be of no consequence, while others consider it to be a major factor in determining the outcome of competition. Broad-spectrum antibiotic production is thought to be a major factor in the antagonistic properties of *B. subtilis*, an important potential biological control agent against fungal pathogens. It seems likely that volatile compounds are also involved (**219–224**).

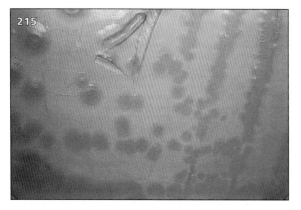

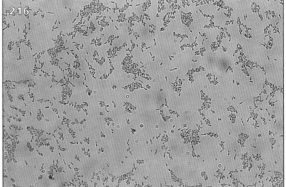

215 *Agrobacterium tumefaciens*. Colonies are unremarkable and either non-pigmented or a light beige-brown. Copious slime is produced on carbohydrate-containing media. (Source: soil around infected seedlings; Medium: nutrient agar; Incubation: 25°C (77°F), 2 days.)

216 *Agrobacterium tumefaciens*. Aggregates which are distinctly star or rosette shaped may be seen but this property appears to be strain dependent. Much longer cells are common in older cultures. (Source: as for **215**; Microscopy: Gram stain, × 1,200.)

217 Tumour caused by *Agrobacterium tumefaciens* on a young tomato plant. Reproduced by courtesy of Prof. D.G. Jones, Department of Agricultural Sciences, University of Wales, Aberystwyth, UK.

218
A. Rhizobium
1. Gram-negative rod-shaped bacterium.
2. Free-living cells of *Rhizobium* are stimulated by proximity to roots of host (leguminous) plants.
3. Infection thread enters root hairs, may be through wound.
4. Tetraploid host cell followed by diploid host cells stimulated to divide repeatedly to form nodule.
5. *Rhizobium* in nodule loses ability to divide, but maintains metabolic activity for extended period.

B. Agrobacterium
1. Gram-negative rod-shaped bacterium.
2. Free-living cells of *Agrobacterium* are stimulated by roots of host.
3. Enters root hair *via* wound (may also enter other parts of plant).
4. Host cells transformed by *T-DNA* from plasmid to stimulate cell division and formation of tumour.
5. After initial period when growth of *Agrobacterium* stimulated by host-produced opines, bacterium loses ability to multiply while remaining metabolically highly active.

218 Comparison of infection with *Agrobacterium* and *Rhizobium*.

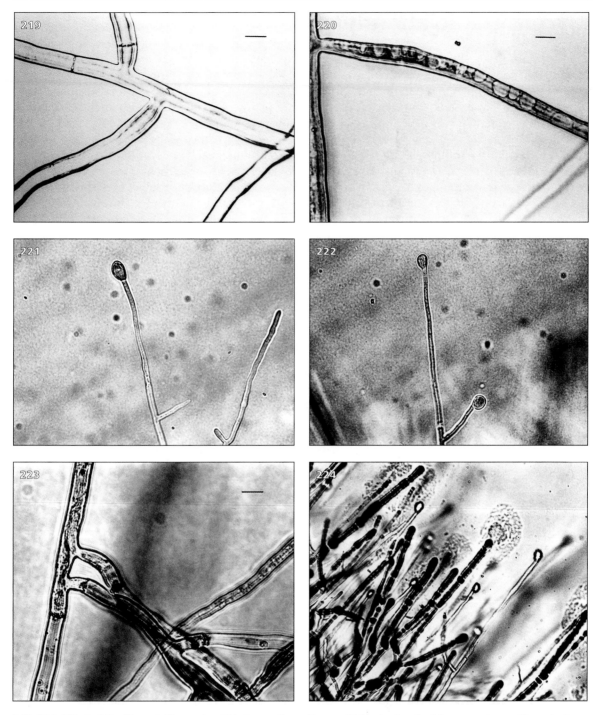

219–224 Effect of *Bacillus* volatiles and antibiotics on the pathogen *Rhizoctonia solani*. In comparison with control hyphae (**219**), volatiles induce vacuolation (**220**) and swelling of the hyphae tip (**221**). Swelling of hyphae tips also occurs in the presence of *B. subtilis* antibiotic (**222**) accompanied by cytoplasmic breakdown within the hyphae (**223**) and lysis of the tips (**224**). **219–224** reproduced with permission from Fiddaman, P.J. and Rossall, S., 1993, © 1993, The Society for Applied Bacteriology.

Mycorrhizal relationships

Mycorrhizal relationships have existed since the Devonian period and some primitive plant orders, such as the Magnoliales, have an obligate dependence on mycorrhizae. Some flowering plant families, however, contain non-mycorrhizal species.

At its simplest, the mycorrhizal relationship is seen as a straightforward mutualistic relationship. The fungi that participate are usually those unable to utilize the complex polysaccharides, which are the principal carbon sources for micro-organisms in many types of soil. Mycorrhizal associations enable these fungi to obtain an abundant supply of simple carbohydrates, such as glucose, from the plant. Fungi of mycorrhizal species may be free-living in soil but generally they have only a limited ability to exist saprophytically.

Infection of plants with mycorrhizal fungi involves a complex interaction involving plant, soil, fungi and other soil micro-organisms. Soil extracts, host root exudate and, specifically, flavonoid compounds are all stimulatory to vesicular–arbuscular (VA) fungi but knowledge of the communication between roots and fungi is incomplete.

The plant benefits from access to nutrients which are otherwise unavailable. This results from the ability of mycorrhizal fungi to hydrolyse large nitrogen- and phosphorus-containing molecules. In addition, hyphae, or an aggregation of hyphae, and fungal cords widen the area of root exploration and are able to bridge zones of low nutrient concentration. Cords are also involved in water transport to the root, thus enhancing drought resistance. Root size and longevity can also be increased by vitamin production by mycorrhizal fungi.

Two broad categories of mycorrhizae exist: ectotrophic and endotrophic. In addition, a small group of ecto/endotrophic mycorrhizae exist, variously termed arbutoid or monotropoid depending on the type of plant partner. Endotrophic mycorrhizae are the most widespread and are further categorized into vesicular–arbuscular (VA), ericoid and orchidaceous. Of these, VA mycorrhizae are most common and, in many cases, are associated with the roots of economically important plants. Endotrophic mycorrhizae are found on about 90% of the world plant flora, though ectotrophic mycorrhizae (**225**, **226**) are of considerable importance in temperate forests. At least 70 fungal species are known to be capable of forming mycorrhizal associations and the actual number is undoubtedly many times greater. With a few exceptions, mycorrhizal fungi are not species

225 An ectomycorrhizal root system developing on the root system of a pine seedling.

226 Fungi forming ectomycorrhizal associations are mainly members of the Agaricales, Hymenomycetes and Basidiomycotina. They form conspicuous fruiting bodies (such as mushrooms and toadstools) on forest floors in autumn and spring.

specific. A given fungus can be associated with several plant hosts while equally a given mycotrophic plant can form associations with one of several fungi. One species of pine, for example, can form associations with any of more than 40 species of mycorrhizal fungi. Despite this lack of specificity, a predominant mycorrhizal type is recognized in each different ecosystem. This is a consequence of Liebig's 'Law of the Minimum' which states that, when all other nutrients are available, the growth of plants is regulated by supplies of that nutrient element that is present in the smallest amount. This 'law' may be extended to heterotrophs such as fungi. Plants growing in different ecosystems face different nutrient limitations, selection favouring the fungi that can most effectively mobilize or capture the limiting nutrient for plant growth in that ecosystem (Read, 1991).

In heathland, and in most boreal and temperate forests, nitrogen is the 'minimal' nutrient and mycorrhizal fungi are able to mobilize that element. Further selection occurs according to the nature of the nitrogen-containing resources. In heathland, nutrients are sequestered in an acidic organic matrix ('mor' humus) that consists of highly recalcitrant residues of the ericaceous plants (e.g. heather) and fungi which are able to grow in this ecosystem. Fungi participating in ericoid mycorrhizae are thus of physiological types capable of hydrolysing lignin, proteins, polyphenols and chitin. In the less severe environments of the boreal and temperate forests, substrates are intermediate in terms of recalcitrance. Plants are of the ectomycorrhizal type, the hydrolytic activity of the participating fungi varying according to the substrate, and being greatest where residues are of high C:N ratio. Only a few ectomycorrhizal fungi are able to break down very complex molecules such chitin but, in general, this group has wider hydrolytic properties than realized previously.

In warmer climates, grasslands and tropical forests develop, the substrates in these exosystems being the least recalcitrant. Phosphorus, rather than nitrogen, is the growth-limiting nutrient and VA mycorrhizae are dominant. In the great majority of cases, VA fungi are of very limited hydrolytic ability and largely unable to mobilize minerals from polymers. However, an extensive hyphal network develops that is highly effective in scavenging soluble phosphate ions. It is also possible that VA fungi are directly involved in the mobilization of phosphorus from recalcitrant sources (Gianinazzi-Pearson and Gianinazzi, 1989).

Although the nutritional aspects of mycorrhizal associations are often considered of greatest importance, the plant may gain other significant advantages. Mycorrhizae may protect the root from pathogens by acting as a mechanical barrier or by secreting antibiotics. Indirect protection may also occur by alterations to the rhizosphere microflora and stimulating competition for ecological niches occupied by pathogens. Mycorrhizal fungi are also of importance in protecting host plants from the toxic effect of pollutants, including acid rain, sulphur dioxide and metal ions, especially heavy metals, aluminium and, at high concentrations, iron. Metal toxicity can be a major problem in soils with low pH value as a consequence of the greater solubility of the metallic ions; ericoid mycorrhizae are often of particular significance. The situation with iron is complicated by the fact that the metal is an essential element at lower concentrations. Under these conditions, ericoid mycorrhizal fungi enhance plant growth by significantly increasing iron uptake. A number of species have been shown to produce hydroxamate siderophores which are probably involved in scavenging for iron when the metal is present only at very low concentrations (Schuler and Haselwandter, 1988).

Mycorrhizal infection is of economic and commercial significance due to the importance in the establishment and growth of plants. In some situations inoculation of plants with the mycorrhizal symbiont is considered necessary (*Table 3*).

Table 3. Applications of mycorrhizal inoculation.

Orchid culture*
Introduction of exotic trees to new sites
Introduction of ericaceous fruit crops
(e.g. blueberry) to new sites
Propagation of woody ornamentals after soil sterilization
Planting of Sitka spruce on first-rotation forest sites
Establishment of trees during remediation of virgin sites, such as coal spoil or waste tips

*Inoculation not now usual commercial practice
Data from Mitchell (1993)

Nitrogen-fixing symbioses

The importance of leguminous plants in crop fertility has been recognized since antiquity. This is a consequence of infection of the plants with nitrogen-fixing rhizobia (*Rhizobium* and *Brady-rhizobium*). The rhizobia are present in variable numbers in normal soil but inoculation of the seed of legumes with pure cultures of rhizobia is a well-established practice to ensure high yield. Alternatively the inoculant may be sprayed into the soil, while dual inoculation with rhizobia and VA mycorrhiza has also attracted attention. Leguminous plants stimulate the growth of rhizobia in the soil to a very considerable extent, effects extending as far as 25 mm from the root surface. Stimulation of free-living cells is highly specific, only *Rhizobium* (**227**) and *Bradyrhizobium* benefiting.

Both *Rhizobium* and *Bradyrhizobium* are highly host-specific. *Bradyrhizobium* is unique in forming a symbiotic relationship with the non-legume *Parasponia*. It has been postulated that *Brady-rhizobium* (which is alkaline-producing and highly suited to life in acid, tropical soils) is the ancestral form from which the acid-producing *Rhizobium* eventually arose, but this is not universally accepted.

Infection of the host plant (**228**) is initiated by penetration of the root hair by a group of *Rhizobium* cells. Invagination of the root hair membrane leads to formation of an infection thread which penetrates the root cortex. Bacterial cells are ultimately liberated into the host cell cytoplasm enclosed in a membrane. The characteristic nodule (**229, 230**) is formed when the infection thread reaches a tetraploid cell of the cortex which, along with neighbouring diploid cells, is stimulated to constant division. Young bacterial cells in the nodules are rod-shaped but later become irregular in shape as metabolically active but non-dividing 'bacteroids'. Nitrogen-fixation continues so

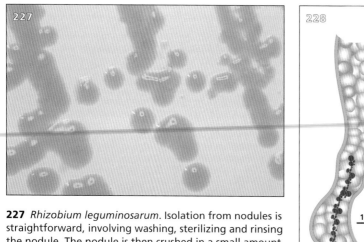

227 *Rhizobium leguminosarum*. Isolation from nodules is straightforward, involving washing, sterilizing and rinsing the nodule. The nodule is then crushed in a small amount of sterile water and the resulting suspension streaked on to solid medium. (Source: nodules, white clover; Medium: yeast extract–mineral salts–mannitol; Incubation: 28°C (82.4°F), 3 days.)

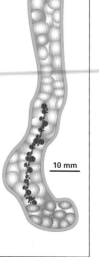

228 Infection of leguminous plants with *Rhizobium*. A diagram of a newly infected root hair shows the bacterial infection thread passing up the hair. Infection results in a progressive curling of the root hair tip.

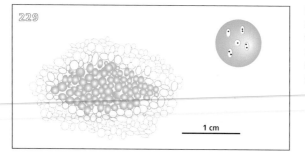

229 A section through a characteristic nodule illustrated diagrammatically. The green cells indicate those filled with bacteria. The magnified part shows one of the mature nodule cells having a large membrane envelope containing four to six 'bacteroids'. No further bacterial growth occurs.

long as the cells are metabolically active. Free-living cells do not fix N_2.

Root nodules are also produced in a nitrogen-fixing association between the actinomycete *Frankia* (**231**) and a very wide range of non-legumes. Nodules are of two types, negatively geotrophic roots in the Casuarinales and Myricales and repeatedly branching roots in other species. The role of cyanobacteria in nitrogen-fixing symbioses is also important although root nodules are not formed (**232**).

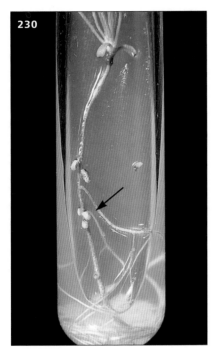

230 *Rhizobium* nodules (arrowed) on the roots of young, white clover. Reproduced by courtesy of Prof. D.G. Jones, Department of Agricultural Sciences, University of Wales, Aberystwyth, UK.

231 Section through a sporangium of *Frankia*. Mature spores are at the apex, young spores developing at the base of the sporangium (x 11,500).

232 The first stage in symbiosis between the cyanobacterium *Nostoc* and liverwort (*Blasia*). Reproduced by courtesy of S. Babic and D.G. Adams, University of Leeds and the Society for General Microbiology.

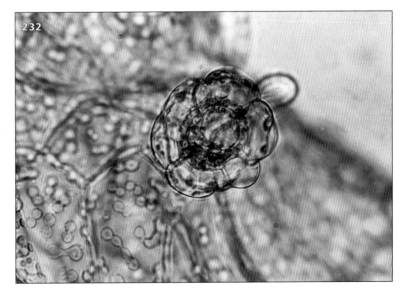

Micro-organisms on the surfaces of plants

The aerial surfaces of plants are colonized by a mixed and characteristic microflora. This is commonly referred to as the epiphytic microflora although this is not an absolute classification since surface micro-organisms may become internal following heavy rain and possibly under other circumstances (Hirano and Upper, 1988). Micro-organisms on plant surfaces as a whole have been placed in two ecological groupings: 'residents' (capable of colonization) and 'casual' (transitory).

Bacteria, yeast and fungi are all present as colonizers of plant surfaces. There is, however, a microbial succession which determines the dominant type. Bacteria are the main early colonists and dominate during the early growing season. Yeasts and yeast-like fungi become of greatest importance during the middle of the growing season and are of overall greatest importance, outnumbering other micro-organisms by up to 50 to 1 (**233–242**).

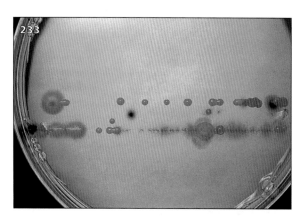

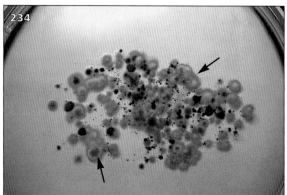

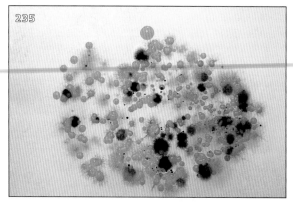

233–235 Epiphytic microflora of pampas-grass (**233**), tragescanthia (**234**) and rose (**235**). All plants were grown in the same area and shared some members of the microflora. Pampas-grass, however, was less heavily colonized while only tragescanthia (grown indoors) was colonized by *Streptomyces* (arrowed; blue-grey colonies superficially resembling moulds).

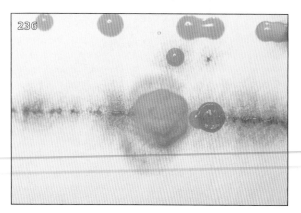

236 The epiphytic microflora of pampas-grass was characterized by slow-growing yeasts and yeast-like fungi which could not be identified with known genera. These are illustrated on a primary isolation plate (see also **237–239**). (Medium: malt extract agar; Incubation: 22°C (71.6°F), 7 days.)

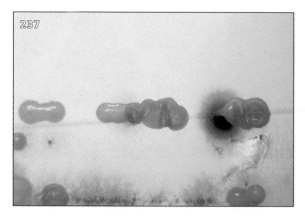

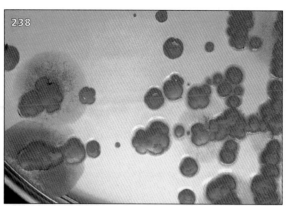

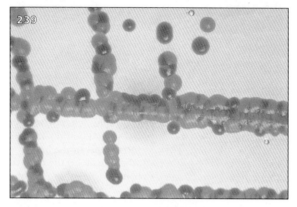

237–239 Slow-growing yeasts and yeast-like fungi from pampas-grass illustrated on a primary isolation plate (**237**) and after initial isolation (**238**, **239**). (Medium: malt extract agar; Incubation: 22°C (71.6°F), 7 days.)

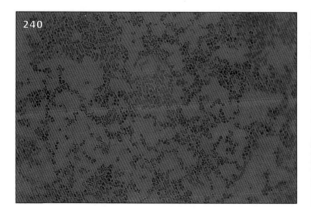

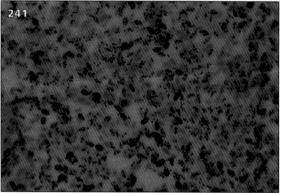

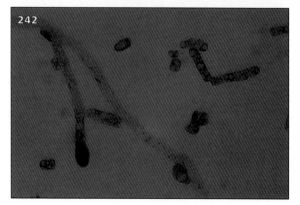

240–242 Photomicrographs of cells of yeasts and yeast-like fungi isolated from pampas-grass. (Microscopy: simple stain; × 40.)

Filamentous fungi follow yeasts and yeast-like fungi (243–247) at the end of the growing season, their hyphae possibly penetrating the tissues and causing minor disease and yield-loss. Microorganisms may also overwinter in dead plant parts (248, 249).

The epiphytic microflora of plants is predominantly saprophytic although many plant

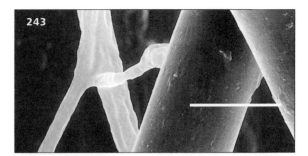

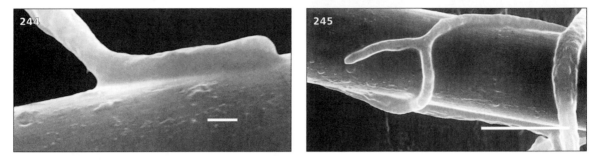

243–245 Scanning electron micrographs demonstrating interactions between *Trichoderma harzianum* and nylon fibres coated with purified lectin from the plant pathogenic fungus *Sclerotium rolfsii*. The first stage (**243**) involves atypical branching of the *Trichoderma* and contact of the branch tip with the fibre surface (bar = 10 μm). This is followed (**244**) by elongation of the firmly attached tip along the fibre surface (bar = 1 μm). The final stage (**245**) involves tight adherence of the *Trichoderma* hyphae to the fibre surface and the production of additional branches (bar = 10 μm).

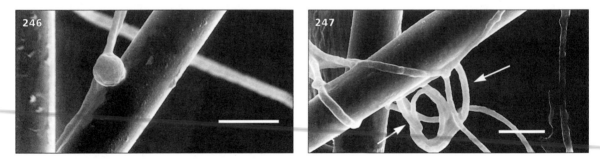

246, 247 Scanning electron micrographs of the mycoparasite-related structures formed by *Trichoderma* during interaction with lectin-treated fibres. Appressorium-like bodies (**246**) are formed at the tip of short branches, while hyphal loops (**247**; arrowed) are distinctive features (bars = 10 μm). **243–247** reproduced by courtesy of Dr J. Fabar, Hebrew University of Jerusalem, Israel.

248, 249 Dead plant material provides a large number of microenvironments which can be of importance in the survival of epiphytic bacteria during winter months. Bracken (**248**) supports a relatively wide range of yeasts while pigmented, Gram-negative bacteria could be isolated from teasel heads (**249**) but not the surrounding material.

pathogens, primarily bacteria, can adopt a resident saprophytic phase on either host or non-host plants. Most of the bacteria present are Gram-negative rods (250, 251) although Gram-positive bacteria may be present as 'casuals'. The most common bacteria are *Erwinia* (252, 253), *Pseudomonas* and *Xanthomonas* (254) together with an ill-defined group usually described as flavobacteria.

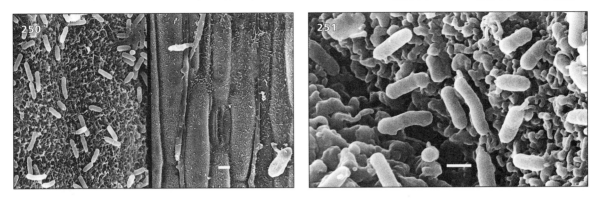

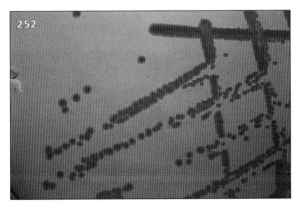
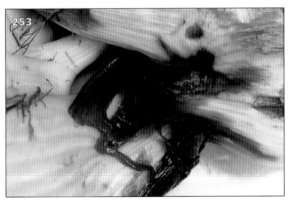

250, 251 Gram-negative, rod-shaped bacteria on the upper (**250**) and lower (**251**) surfaces of barley leaves. Fungal spores and hyphae are also present on the upper surface (bar = 10 μm in **250** and 1 μm in **251**). Reproduced with permission from Blakeman, J.P., 1991, © 1991, Society for Applied Bacteriology.

252 *Erwinia carotovora*. Although most strains of *E. carotovora* are lactose positive, lac⁻ variants are common. Although *E. carotovora* grows well on most common media for the *Enterobacteriaceae*, there is usually no means of differentiation from other members of the family. (Source: infected carrots; Medium: MacConkey purple agar; Incubation: 30°C (86°F), 2 days.)

253 Soft rot lesion on fennel caused by *Erwinia carotovora*. In addition to its role as a plant pathogen, *E. carotovora* is an important cause of spoilage of a wide range of vegetables. The plant may be infected before harvest or the organism may gain entry through post-harvest damage, tissue being subsequently broken down by pectinolytic enzymes. A variety of other micro-organisms may be found in the rotting part but usually have no causal role.

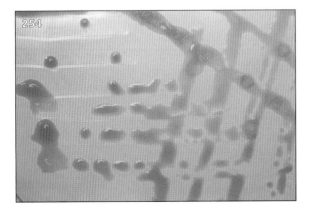

254 *Xanthomonas campestris*. When grown on medium containing high levels of usable carbohydrates, the colonies are mucoid or slimy due to the production of extracellular polysaccharides (xanthan gums). Most, but not all, isolates are yellow-pigmented, the pigments being highly characteristic xanthomonadins (brominated aryl polyenes). Isolation is best made from infected plant material, nutrient agar being satisfactory for most species. (Source: infected pea plants – *Pisum sativum*; Medium: nutrient agar + 5% sucrose; Incubation: 25°C (77°F), 3 days.)

Pathogens with an epiphytic resident phase include the economically important *Erw. amylovora* (fireblight) and *Erw. carotovora atroseptica* (blackleg and tuber rot of potato). The epiphytic microflora is also a source of ice nucleation active (INA) bacteria. Specialized pathogens (which act to form nuclei ice crystals) cause, usually localized, damage to plants during frost. The most common yeasts are *Aureobasidium pullulans*, *Cryptococcus* spp. (**255**), *Rhodotorula*, spp. and *Sporobolomyces* spp.

Life on the surfaces of plants is rigorous and provides higher levels of environmental stress than other plant-associated habitats such as roots. Water is usually available only intermittently and cells are subject to damage by UV irradiation. In hot climates, surface temperatures may be sufficiently high to damage more sensitive cells. Micro-organisms are also subject to inhibition by plant-produced inhibitors such as phytoalexins; levels of pollutants may also be high. Nutrients are derived from leachates, originating within the plant. Considerable competition exists between epiphytic micro-organisms for niches on the plant surface. A mechanism known as exclusion is operative in which the first bacterium to arrive at a niche is able to exclude other bacteria, often closely related, with the same site preference (Blakeman, 1991). A number of competitive strategies may be used, including both antibiotic and bacteriocin production. In general, antibiotic production is limited to a small proportion of the epiphytic microflora. *Pseudomonas fluorescens*, however, produces both antifungal and antibacterial antibiotics and *in vivo* inhibition of the pathogen *Ps. syringae* pv. *phaseolicola* can be demonstrated.

Yeast and yeast-like fungi were not previously thought to produce antibiotics, but production has now been demonstrated (McCormack *et al.*, 1994). Strains of both *Erwinia* and *Pseudomonas* have been shown to produce bacteriocins against closely related bacteria. The *in vivo* significance of bacteriocin production is difficult to evaluate in many situations but there is some evidence of involvement of bacteriocins in the inhibition of *Erw. amylovora* by *Erw. herbicola*.

Despite antibiotic and bacteriocin production by plant epiphytes there is evidence that at least part of the basis for competition is nutritional. This is likely to involve specific nutrients. There is strong nutritional competition between, for example, saprophytic bacteria and necrotrophic fungal pathogens such as *Botrytis cinerea* and *Phoma betae*. The outcome is that amino acids become limiting, with adverse effects on the ability of the fungi to germinate (Blakeman and Brodie, 1977). The exclusion mechanism may also be explained, at least in part, by the enhanced ability of bacteria established in niches to obtain nutrients. Antibiotic production and nutritional limitation may, however, be linked. In pure culture, antibiotic production by yeast was greatest when nutrients were limiting (McCormack *et al.*, 1994). This suggests that antibiotic production is of importance in enhancing the ability to compete for nutrients.

Attempts have been made to exploit the exclusion mechanism to prevent pathogens from becoming established as part of the epiphytic flora. Some success has been achieved in excluding INA bacteria with either mutant INA-negative strains or artificially constructed, isogenic, INA-negative strains. Experimental work is also being conducted in attempts to control *Ps. syringae*.

Some of the epiphytic bacteria are involved in the breakdown of large masses of vegetation after its death. Numbers of *Enterobacteriaceae* increase initially, followed by lactic acid bacteria as the pH value falls and micro-aerophilic conditions develop. Natural heating can mean a rapid temperature rise and the growth of thermophilic bacteria (see 'Landfill', page 122). Large shrubs and trees break down only slowly and provide a wide range of habitats for micro-organisms (**256**) and higher life forms, including small mammals. Micro-organisms involved are primarily fungi (**257**, **258**) although slime moulds (Myxomycetes; **259–262**) and bacteria are also present.

255 A yeast of atypical colonial morphology resembling *Cryptococcus*. The yeast was isolated from beech leaves in several parts of a forest during late summer but not from surrounding soil. (Medium: oxytetracycline– glucose–yeast extract agar; Incubation: 22°C (71.6°F), 5 days.)

256 Colonization of a fallen tree trunk. A tree trunk, which had fallen into a semi-waterlogged area, was colonized by a remarkable variety of micro-organisms including bacteria, microfungi and protozoa. Nematodes were also present in large numbers. No systematic evaluation of bacterial species was made but diversity was considerable. 'Colonies' developing in the slime matrix were polymicrobic aggregations comprising morphologically distinct bacteria and, possibly, yeast.

257, **258** Fruiting bodies of higher fungi are common on decaying wood as well as on live trees. The presence on decaying wood does not necessarily mean that the fungus is of a wood-rotting species.

259–262 Slime moulds (Myxomycetes) are common on decaying wood including wood used for constructional purposes.

The lichen symbiosis

Lichens are composite organisms consisting of a fungal partner (which is usually dominant) and one or more photosynthetic partners (**263**). The composite structure is generally considered to behave as an independent entity. The fungal partner, the mycobiont, is usually an ascomycete, although basidiomycetes are occasionally lichenized. The photosynthetic partner, the photobiont, may either be a green alga or a cyanobacterium. Some lichen symbioses involve three, four or more partners. Although the mycobiont is dominant in terms of providing the bulk of the tissue, the success and longevity of the symbiosis ultimately depends on the photobiont. Photobionts come from a small number of genera, generally widespread in nature, while mycobionts are taxonomically diverse but exclusively lichen forming (*Feature 11*).

From an ecological viewpoint, the lichen symbiosis is considered to be clearly mutualistic since it enables both partners to colonize areas where neither could survive alone. The association may therefore be described as obligate in that the '… fitness of both partners approaches zero when they are not together' (Lewis, 1987). From a purely mechanistic viewpoint, the mycobiont is dependent on the photobiont for organic nutrients. The photobiont is nutritionally independent of the mycobiont but appears to benefit by the enhanced uptake of

water and minerals and protection from desiccation. Parasitic lichens, of the genus *Rhizocarpon*, are also recognized – these are parasitic on other lichens, including other *Rhizocarpon* species (Poelt, 1991).

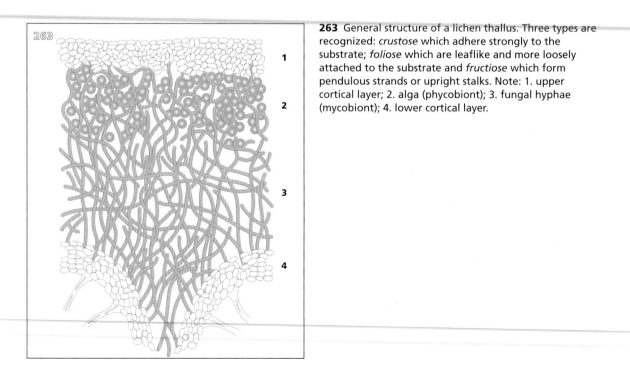

263 General structure of a lichen thallus. Three types are recognized: *crustose* which adhere strongly to the substrate; *foliose* which are leaflike and more loosely attached to the substrate and *fructiose* which form pendulous strands or upright stalks. Note: 1. upper cortical layer; 2. alga (phycobiont); 3. fungal hyphae (mycobiont); 4. lower cortical layer.

264–266 *Et in Arcadia ego.* Lichen-encrusted tombstones are an evocative sight in a country churchyard (**264**). Most on stone in the UK are of the crustose type (**265**). Lichens are highly adapted to their microenvironment and apparently minor changes caused, for example, by a tombstone tilting with age, can upset the balance between the lichen and its surroundings and lead to death (Hawksworth, 1990). Lichens are sensitive to heavy metals and are unable to develop on tombstones where lead lettering is used (**266**).

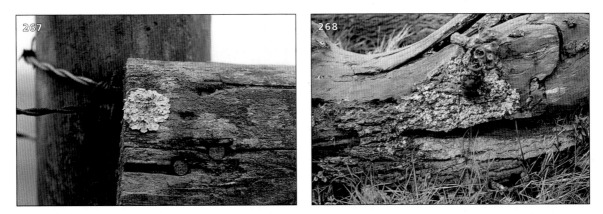

267, 268 A solitary foliose lichen (**267**) on a fence post representing successful colonization of a microhabitat. A few feet away more favourable conditions on a fallen tree (**268**) permit wider colonization.

From a global viewpoint, lichens are significant in a number of ways. As carbon sinks, lichens fix vast but unquantified amounts of carbon through photosynthesis. Triterpenoids, produced in large quantities by large, rainforest lichens, may contribute to the pool of buried carbon, usually considered to consist of recalcitrant compounds. Such triterpenoids probably have an anti-herbivore function and prevent the utilization of rainforest lichens as a protein source by herbivores. Lichens, however, also have a significant role in nutrient cycling and the food web.

Relatively fast-growing lichens, which have a short life-cycle on short-lived substrates, are of importance in tropical environments. In most cases the growth of lichens is slow and can continue for centuries. Competition is rarely a problem as lichens develop in continuous equilibrium with their immediate environment though the balance is delicate and highly susceptible to disturbance by environmental change. This very susceptibility, however, means that lichens are of considerable value as environmental monitors. Disturbance affects the presence or absence of species and may also involve reduction in frequency, luxuriance, fertility and the ability to establish new colonies (Hawksworth *et al.*, 1974). Lichens are most widely used as biomonitors in studying air pollution and are particularly sensitive to sulphur dioxide (**269**). Lichens have also been used to measure other atmospheric pollutants (Hawksworth, 1990; Richardson, 1991) and, on a much wider scale, climatic change (*Table 4*; Galloway, 1992; Smith, 1990).

Man also makes some direct use of lichens, including medicines (**270**), foods, dyes and as perfume ingredients (*Table 5*). Lichens also interact with man in other ways: although none is a recognized pathogen, secondary metabolites are responsible for contact dermatitis amongst forestry workers (woodcutter's eczema). Allergic responses to perfumes and aftershave lotions are often also associated with the use of lichen extracts (Richardson, 1991).

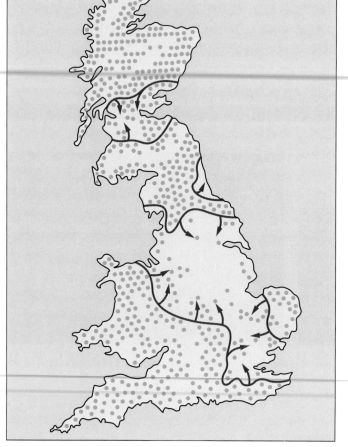

269 Changes in distribution of *Usnea* in mainland Britain following the implementation of the Clean Air Acts. Solid lines indicate areas where lichens disappeared during 1800–1970. Arrows indicate the advance of recolonization.

Table 4. Use of lichens in environmental monitoring.

Environmental area	Monitoring for
Air pollutants	Sulphur dioxide
	Heavy metals
	Radionuclides
	Fluorides
	Acid rain
Climatic changes	Global warming
	Ozone thinning
	(increased UV radiation)
	Dating of rock surfaces

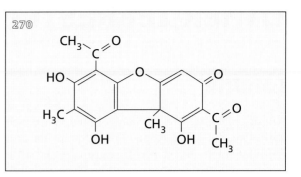

270 Structure of usnic acid, a lichen secondary metabolite (lichen acid). This compound was previously used as a topical antibiotic.

Table 5. Uses of lichens by man.

Area of use	Use
Medicine	Topical antibiotics such as usnic acid (Usno®)*
	Possible future antibiotics
	Use of Iceland moss *Cetraria islandica* in home-made tonics and laxatives
	Use of *Cetraria islandica* or *Usnea* spp. in commercial remedies for sore throats
Food	Traditional food in times of hardship
	Japanese delicacy prepared from *Umbilicaria esculenta*
	Curry additive Garam masala contains a high proportion of various lichens
Perfumes and toiletries	Use of *Evernia prunastri* (oakmoss) and *Pseudevernia furfuraceae* (treemoss) as fragrance and fixative in aftershaves, perfumes and soaps
Miscellaneous	Use of *Cladonia stellaris* in wreaths, floral decorations and architects' models
	Manufacture of the acid–base indicator litmus from *Roccella* spp.

*Commercial production now ceased due to uncertainty of raw material supply
Information from Richardson (1991)

OTHER TERRESTRIAL HABITATS

Landfill

Municipal waste, which includes domestic rubbish and small, but variable, quantities of industrial and building waste, is usually disposed of in landfill. Landfill sites may also receive various types of clinical waste. The most acceptable approach is the sanitary or engineered site, where waste is covered with soil and compacted immediately after deposition. Leachate is contained and treated and the site is managed to minimize nuisance and avoid pollution.

A very wide variety of materials are present in landfill and a correspondingly wide variety of degradative processes occur. Three major processes are involved: decomposition of cellulose-based materials to sugars, formation of weak acids from the sugars and methanogenesis. Open sites are initially aerobic but sanitary sites become anaerobic very quickly. In either case a temperature rise to $c.60°C$ ($140°F$) is common and encourages thermophilic micro-organisms including species of *Bacillus* and, at suitable oxygen tensions, the mycelial *Thermoactinomyces*.

Very considerable methanogenesis occurs in landfill. In some cases sanitary landfill sites have been constructed to allow the recovery of methane and its subsequent use as fuel. In many cases, however, methane is a cause of considerable problems, leakage from sites being a nuisance and contributing to the greenhouse effect. Methane seeping from old landfill sites can cause problems in modern buildings erected many years after use of the site has ceased, hence modern site management often involves capping the site with an impermeable clay coat; this procedure is expensive and not always effective. A possible alternative and more economical management strategy is to maintain an open, permeable soil cover and manipulate conditions to optimize the microbial oxidation of methane. In the longer term, however, a better policy is to minimize use of landfill (*Feature 12*).

In the UK, considerable publicity has been given to the finding of clinical material, as well as blood- or faeces-stained material at landfill sites. This has raised fears of environmental contamination with pathogens. Evaluations have shown that the survival of many pathogens in landfill is limited. Some movement in leachate has been reported but, in general, the risk is considered to be low (Collins and Kennedy, 1992).

Lithobiotic (rock-inhabiting) habitats

Lithobiotic micro-organisms are widespread in nature although relatively little is known of their ecology. They colonize those exposed rock surfaces that are not soil covered or occupied by more aggressive higher organisms such as mosses or sessile animals. A wide range of climates and habitats support the growth of lithobionts ranging from both hot and cold deserts to submerged marine and freshwater rocks, and from deep caves to the surface of buildings (Friedmann and Ocampo-Friedmann, 1984).

There is evidence for involvement by micro-organisms with cyanobacteria, acid-producing fungi, thiobacilli and nitrifying bacteria being particular candidates (271–273). Testate protozoa (*Difflugia* and *Euglypha* spp.) may also be present. Microbial degradation of the built environment is not, of course, restricted to stone (274).

Feature 12. So where does it all go?

Despite the widespread assumption for an anaerobic route of mineralisation in landfill, and the observed quantities of methane evolved, the amount of the gas is usually less than 10% of that theoretically predicted. Recognising and understanding the limitations of methanogenesis in landfill is seen as being of increasing importance. In some cases at least, low rates of methanogenesis can be directly related to the thermodynamic uncoupling of fatty acid oxidation although underlying causes are probably low pH values and an accumulation of toxic, volatile, fatty acids. Data from Mormile *et al.*, 1996.

271–273 Deterioration of building stone. Many factors are involved and chemical pollution is generally thought more significant. It is, however, considered that micro-organisms have a strong influence when cell numbers reach 10^4 to 10^6 g^{-1}. Corroded sandstone (**271**) bore a mixed bacterial and yeast microflora, but the role in corrosion is not known. Adjacent brickwork (**272**) was contaminated with low numbers of a species of *Bacillus*. This organism had a characteristic colonial morphology and was alkali resistant (**273**). The *Bacillus* was probably a secondary invader. The organism could not be isolated from surrounding soil although a morphologically similar *Bacillus* was isolated from 'sarsen stones' at Fyfield Down, Wiltshire, UK. Again, attempts at isolation from surrounding soil failed. (The sarsen stones at Fyfield Down are the remains of a limestone cap which once covered southern England. The down was the source of stones used to build the great stone circles at nearby Avebury and has been described as the finest prehistoric landscape in Europe.)

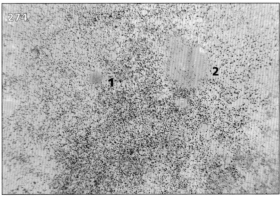

274 Microbial colonization of interior walls. The growth of mould on interior walls is commonplace and usually associated with condensation or penetrating moisture. In this instance the affected wall showed little sign of mould growth although the paint was badly discoloured and the wall 'sticky' to touch. Culture of swabs taken from the wall showed heavy mixed growth of mould, together with bacteria (1) and yeast (2).

From the viewpoint of man, lithobiotic micro-organisms occupy desolate and hostile environments but it is important to distinguish between conditions in the macroclimate and those immediately surrounding the microbial communities; the term 'nanoclimate' has been proposed for the latter. Under most circumstances rock must be regarded as an extreme environment. Even in the humid north European climate, stone as a biotope offers conditions for organisms similar to those of deserts and semi-deserts (Truper and Galinski, 1986). Lithobiotic micro-organisms are subject to extremes of temperature, desiccation and, at times, of low moisture content, salt stress due to the dissolved salt concentration increasing to crystallization.

Lithobiotic bacteria, and possibly other micro-organisms, possess both an external and an internal cellular protection system. External protection is offered by life in biofilms. The biopolymer matrix itself conserves water through its nature as a hydrogel but a more specific mechanism is the accumulation, in microniches, of hygroscopic metabolites such as $NaNO_3$ which form 'water nests'. Water thus remains available during drought conditions, permitting the survival of bacteria and other micro-organisms sensitive to total dehydration (Moser et al., 1990). Internal protection is offered by synthesis of protective substances such as mono- and disaccharides, amino acids and betaines, which minimize the effect of salt stress (Bock and Sand, 1993).

Three main categories of lithobionts have been proposed although terms are not absolute and many types are best described as intermediate: hypoliths reside under pebbles and small stones on soil (they are restricted to hot desert pavements and are largely cyanobacteria, green algae and heterotrophic bacteria), endoliths live inside rocks, while epiliths live on the surface.

Endolithic micro-organisms are of particular note in their ability to exploit relatively favourable nano-climates within the rock, contrasting markedly with the abiotic interior. Endoliths have been further classified into euendoliths, chasmoendoliths and cryptoendoliths. Euendoliths, which actively penetrate the rock fabric, are absent from extreme arid environments, probably because penetration requires a minimum water content. Euendolithic micro-organisms impinge directly on man in that the exterior surfaces of buildings, statues, and the like, are among the substrates colonized. Constructional stone inevitably undergoes physical and chemical deterioration, the rate of which has been increased by air pollution, especially acid rain. Chasmoendoliths dwell within fissures and cracks, which provide sufficient space for microbial growth. Only certain rock types provide a suitable substrate. These are usually weathered rocks such as granite, calcite, quartzite and dolomite. As with other types of endolithic communities, only light-coloured, or translucent, rock permits sufficient light penetration for photosynthetic micro-organisms which are invariably part of the community. Chasmoendolithic environments permit microbial life in both hot and cold extreme arid areas.

Cryptoendoliths colonize porous rocks, such as sandstone, which have sufficient air space to permit microbial growth. The morphology of colonization is little affected by the macro-environment. The outer 1–3 mm of the rock forms a crust in which no micro-organisms develop. Colonies develop in the next few millimetres, adhering to or growing between crystals. The airspace is thus sealed from the exterior by the outer crust of rock. This crust is of considerable ecological importance in insulating the internal airspace system from the external climatic extremes. In the case of Antarctic deserts, this means that lichens require no unusual physiological adaptation though morphogenetic adaptation to a filamentous form is required to permit the lichen to penetrate the porous substrate. Lichens are found only in cold desert environments, these organisms, and indeed all eukaryotes, appear unable to cope with the environmental stresses associated with endolithic life in hot, extremely arid, deserts. Cryptoendolithic micro-organisms inhabit a 'closed ecosystem' in which a steady state exists where photosynthetic activity is balanced by respiration. It has been stated that cryptoendolithic communities are among the most highly closed ecosystems in nature (Friedmann and Ocampo-Friedmann, 1984).

Epilithic life is not restricted to the terrestrial environment, such micro-organisms colonizing rocks in both fresh and marine waters. Epilithic micro-organisms are found in humid conditions in caves and are also involved in the colonization of masonry. Epilithic micro-organisms are associated with the formation of 'desert varnish' on rocks.

Deep, subsurface communities of micro-organisms

For practical reasons there have been relatively few studies of micro-organisms in deep, subsurface habitats although it is well established that significant populations exist at depths in excess of 400 m. The size and nature of populations is dictated by the nature of the substrata but, in general, communities are diverse and metabolically active, consisting of aerobic, micro-aerophilic and anaerobic micro-organisms. These micro-organisms are identical, or closely related, to their counterparts at the terrestrial surface. Activities are greatest in water-saturated, subsurface sands and are least in relatively impermeable clay zones. Methophils are particularly numerous in some groundwaters where optimal conditions for growth occur due to the presence of methane diffusing from anaerobic sediments, where methanogenic activity can be considerable, and of oxygen diffusing from the atmosphere. Methophils may, however, be sensitive to common pollutants such as halogenated aliphatic hydrocarbons (e.g. dichloromethane). Metabolically active, sulphate-reducing bacteria have also been isolated from deep, subsurface sediments despite the low nutrient levels generally assumed to be present. Sulphate-reducing bacteria in sandy aquifer sediments in the London basin were able to utilize a variety of organic substrates while sulphate reduction was supported by an indigenous organic source within the sediments. In this situation, however, sulphate reduction will be limited by the presence of O_2, the quantity and quality of organic carbon available and, possibly, SO_2. Sulphate-reducing bacteria may play a beneficial role in the degradation of recalcitrant organic compounds. There is a limited effect on groundwater quality through small increases in iron concentration (Johnson and Wood, 1993).

The limitation of bacterial populations occurs by protozoal grazing, as in surface soils, and it is presumed that bacteriophages are also present. Anaerobic protozoa are present in anoxic sediments, usually being members of the Archezoa. Anaerobic, free-living, ciliated protozoa (275–277), including species of *Caenomorpha*, *Metopus* and

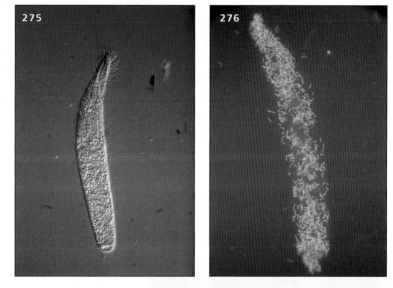

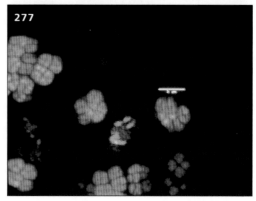

275–277 Free-living, anaerobic, ciliated protozoa and symbiotic methanogens. A living cell of *Metopus palaeformis* (**275**, x 290) alongside the same ciliate (**276**, x 390) showing the presence of autofluorescing methanogens. A confocal image of a dual-probed (bacteria/Archeae) ciliate, *Cyclidium porcatum* (**277**, x 570), illustrates a new type of symbiosis. Bacteria fluoresce green and Archeae (methanogens) red within the intracellular tripartite complex. The red-fluorescing clusters surrounding the ciliate are *Methanosarcina*. Reproduced with permission from Embley, T.M. and Finlay, B., 1994, © 1994, The Society for General Microbiology.

Sanderia, have been isolated from anaerobic, sulphide-rich sands (Fenchel *et al.*, 1977). These protozoa contain methanogenic bacteria as endosymbionts. Mitochondria and cytochrome oxidase are lacking but hydrogenosomes, specialized redox organelles present in other anaerobic protozoa and fungi, are present. Hydrogenosomes oxidize pyruvate with the production of ATP, CO_2, H_2 and acetate. This H_2-evolving fermentation appears to be the basis of the symbiosis, methanogens being shielded from competition with sulphate-reducing bacteria (*Feature 13*).

Feature 13. Deep six

Deep, subsurface micro-organisms are of considerable potential importance in the bioremediation of terrestrial sediments and aquifers polluted by organic and radioactive wastes. The populations, as a whole, have considerable metabolic versatility and are able to degrade a wide range of organic compounds including hydrocarbons, aromatic compounds and chlorinated aliphatic compounds. The latter category includes compounds, such as trichlorethylene, that are highly recalcitrant and serious environmental pollutants. Methanogens can metabolize a range of pollutants and are of importance in this role in a number of environments. Studies of consortia degrading toluene (**278**) and o-xylene (**279**) showed a high degree of substrate specificity and a long adaptation period. Pollutants may not be degraded in the presence of more readily metabolized co-contaminants. DNA repair mechanisms are highly conserved in bacteria in deep, subsurface communities and appear to be of importance in maintaining the integrity of DNA and providing protection against chemical insults, such as oxygen radicals, during periods of slow growth. From Arrage *et al.*, 1993.

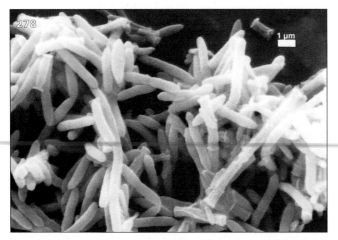

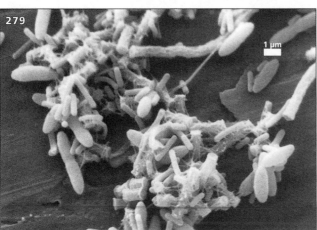

278, 279 A toluene-degrading methanogenic consortium (**278**) was composed primarily of two distinct rod-shaped bacteria, one with rounded ends, the other with blunt ends abutting on to the next cell in the chain. These latter bacteria appeared similar to *Methanothrix soehngenii*. An o-xylene-degrading consortium (**279**) had similar constituents although the round-ended rods were fatter and more ellipsoidal. A thin rod was also more prevalent and a notable feature of the consortium was the web of exopolysaccharide-like material. Reproduced with permission from Edwards, E.A. and Grbic-Galic, D., 1994. © 1994 American Society for Microbiology.

Metal ores

Some micro-organisms, including bacteria, algae and lichens, are able to live in ores containing high metal concentrations. In some cases, the organism is highly tolerant of high concentrations of metals but has no other interaction with the metal. This situation tends to favour microfungi and the actinomycetes. In many cases the metal is solubilized as a consequence of microbial metabolism. Solubilization may be beneficial both to the active micro-organism and to other organisms due to the mobilization of nutritionally important metals. In some cases immobilization leads to an increase in concentration of metals to a toxic level.

An unusual relationship occurs between penicillin-producing fungi and *Bacillus cereus* in areas containing high metal concentrations. Penicillin-producing fungi are highly resistant as a consequence of the breakdown of penicillin into the metal-complexing substance penicillamine. *Bacillus cereus* is not inherently resistant to metals but is highly penicillin resistant and is able to use the fungal defence mechanism against metal toxicity.

Solubilization may be direct (due to enzyme-mediated activity) or indirect (due to the production of corrosive metabolites such as acids, ligands and NH_3). Direct attack involves either oxidation or reduction of the metal and is usually beneficial to the micro-organism. These can only be prokaryotes since the electron transport system is located in the cell envelope and not in the intracellular organelles. Oxidative reactions provide a partial or total source of energy and are themselves of two types, direct or indirect. Oxidative reactions are beneficial in that the mineral serves as either a partial or complete energy source. In reducing processes, the mineral serves, to a greater or lesser extent, as the electron sink for respiration.

Direct attack on ores is of considerable economic importance in that the process can be exploited as a means of extracting minerals from low-grade mined materials and also in mining itself.

Indirect attack on metal ores occurs when metabolic products react with an ore and cause eventual solubilization. This involves several possible mechanisms (Ehrlich, 1992).

A number of micro-organisms also have sequestering properties for metals in solution. These are important in an overall sense in that sequestration may be the nucleation process for *de novo* mineral deposition in nature. Sequestration is also important as a means of removing toxic metal waste, including radioactive metals, from solutions. The process is also involved in the flotation method for the recovery of metals from ores. Micro-organisms may also be used as a means of locating ore deposits. The basis is to search for micro-organisms known to be associated with ore deposits. A number of approaches are possible (*Table 6*).

Table 6. A miner forty-niner – use of micro-organisms in prospecting for metal ores.

Enrichment methods for detection of index organisms
Use of DNA probes for detection of index organisms
Detection of bacteria with unusually high resistance to ores sought
Survey for high concentration of *B. cereus*

REFERENCES

Ambus, P. (1993) Control of denitrification enzyme activity in a streamside soil, *FEMS Microbiology Ecology*, **10**, 225–34.

Arrage, A.A., Phelps, T.J., Benoit, R.E. and White, D.C. (1993) Survival of subsurface microorganisms exposed to ultraviolet irradiation and hydrogen peroxide, *Applied and Environmental Microbiology*, **59**, 3345–50.

Astrom, B., Gustafsson, A. and Gerhardson, B. (1993) Characteristics of a plant deleterious rhizosphere pseudomonad and its inhibitory metabolite(s), *Journal of Applied Bacteriology*, **74**, 20–8.

Blakeman, J.P. (1991) Foliar bacterial pathogens: epiphytic growth and interactions on leaves, *Journal of Applied Bacteriology (Supplement)*, **70**, 49–59.

Blakeman, J.P. and Brodie, I.D.S. (1977) Competition for nutrients between epiphytic micro-organisms and germination of spores of plant pathogens on beetroot leaves, *Physiological Plant Pathology*, **10**, 29–42.

Bock, E. and Sand, W. (1993) The microbiology of masonry deterioration, *Journal of Applied Bacteriology*, **74**, 503–14.

Carter, J.P., Hsiao, Y.H., Spiro, S. and Richardson, D.J. (1995) Soil and sediment bacteria capable of aerobic nitrogen respiration, *Applied and Environmental Bacteriology*, **61**, 2852–8.

Collins, C.H. and Kennedy, D.A. (1992) The microbiological hazards of municipal and clinical wastes, *Journal of Applied Bacteriology*, **73**, 1–6.

Edwards, E.A. and Grbic-Galic, D. (1994) Anaerobic digestion of toluene and o-xylene by a methanogenic consortium, *Applied and Environmental Microbiology*, **60**, 313–22.

Ehrlich, H.L. (1990) *Geomicrobiology*, Marcel Dekker, New York.

Ehrlich, H.L. (1992) Metal extraction and ore discovery, *Encyclopedia of Microbiology* (ed. J. Lederberg), pp. 75–80, Academic Press, New York.

Embley, T.M. and Finlay, B. (1994) The use of small subunit rRNA sequences to unravel the relationship between anaerobic ciliates and their methanogenic endosymbionts, *Microbiology*, **140**, 225–35.

Evans, G.M. (1993) The use of microorganisms in plant breeding, *Exploitation of Microorganisms* (ed. D.G. Jones), pp. 225–48, Chapman & Hall, London.

Fenchel, T., Perry, T. and Thane, A. (1977) Anaerobiosis and symbiosis with bacteria in free-living ciliates, *Journal of Protozoology*, **24**, 154–63.

Fetzer, S., Bak, F. and Conrad, R. (1993) Protective role of pyrites (FeS_2) and survival of methanogenic bacteria, *FEMS Microbiology Ecology*, **12**, 107–115.

Fiddaman, P.J. and Rossall, S. (1993) The production of antifungal volatiles by *Bacillus subtilis*, *Journal of Applied Bacteriology*, **74**, 119–26.

Friedmann, E.I. and Ocampo-Friedmann, R. (1984) Endolithic microorganisms in extreme dry environments: analysis of a lithobiontic microbial habitat, *Current Perspectives in Microbial Ecology* (ed. C.A. Reddy), pp. 177–85, American Society for Microbiology, Washington DC.

Galloway, D.J. (1992) Biodiversity: a lichenological perspective, *Biodiversity and Conservation*, **1**, 312–23.

Gianinazzi-Pearson, V. and Gianinazzi, S. (1989) Phosphorus metabolism in mycorrhizas, *Nitrogen, Phosphorus and Sulphur Utilisation by Fungi* (eds. L. Boddy, R. Marchant and D.J. Read), pp. 227–43, CUP, Cambridge.

Hawksworth, D.L. (1988) Coevolution of fungi with algae and cyanobacteria in lichen symbioses, *Coevolution of Fungi with Plants and Animals* (eds. K.A. Pirozynski and D.L. Hawksworth), pp. 125–48, Academic Press, London.

Hawksworth, D.L. (1990) The long-term effects of air pollutants on lichen communities in Europe and North America, *The Earth in Transition: Patterns and Processes of Biotic Impoverishment* (ed. G.M. Woodwell), pp. 45–64, CUP, Cambridge.

Hawksworth, D.L., Coppins, B.J. and Rose, F. (1974) Changes in the British lichen flora, *The Changing Flora and Fauna of Britain* (ed. D.L. Hawksworth), pp. 47–78, Academic Press, London.

Hirano, S.S. and Upper, C.D. (1983) Ecology and epidemiology of foliar bacterial plant pathogens, *Annual Review of Phytopathology*, **21**, 243–69.

Hirano, S.S. and Upper, C.D. (1988) Rain and the population dynamics of *Pseudomonas syringae* on snap bean leaflets, *Proceedings of the 5th International Congress of Plant Pathology*, p. 86, Kyoto.

Johnson, A.C. and Wood, M. (1993) Sulphate-reducing bacteria in deep aquifer sediments of the London basin: their role in anaerobic mineralization of organic matter, *Journal of Applied Bacteriology*, **75**, 190–7.

Jordan, D.C. (1984) *Rhizobium, Bergey's Manual of Systematic Bacteriology*, Vol. 1 (eds. N.R. Krieg and J.G. Holt), pp. 235–42, Williams and Wilkins, Baltimore.

Kusel, K. and Drake, H.L. (1995) Effects of environmental parameters on the formation and turnover of acetate by forest soils, *Applied and Environmental Microbiology*, **61**, 3667–75.

Lewis, D.H. (1987) Evolutionary aspects of mutualistic associations between fungi and photosynthetic organisms, *Evolutionary Biology of the Fungi* (eds. A.D.M. Rayner, C.M. Brasier and D. Moore), pp. 161–78, CUP, Cambridge.

Marsh, P., Toth, I.K., Meijer, M. *et al.* (1993) Survival of the temperate actinophage OC31 of *Streptomyces lividans* in soil and the effects of competition and selection on lysogeny, *FEMS Microbiology Ecology*, **13**, 13–22.

McCormack, P.J., Wildman, H.G. and Jeffries, P. (1994) Production of antibacterial compounds by phyllophane-inhabiting yeasts and yeastlike fungi, *Applied and Environmental Microbiology*, **60**, 927–31.

McCurdy, H.D. (1986) Myxococcales, *Bergey's Manual of Systematic Bacteriology*, Vol. 3 (eds. P.H.A. Sneath, N.S. Mair, M.E. Sharpe and J.G. Holt), pp. 2139–2148, Williams and Wilkins, Baltimore.

Mitchell, D.T. (1993) Mycorrhizal associations, *Exploitation of Microorganisms* (ed. D.G. Jones), pp. 169–96, Chapman & Hall, London.

Mormile, M.R., Gurijala, K.R., Robinson, J.A. *et al.* (1996) The importance of hydrogen in landfill fermentations, *Applied and Environmental Microbiology*, **62**, 1583–8.

Moser, E., Goretzki, L. and Bock, E. (1990) Hohe stickstoffdeposition an gebaudeoberflachen in Wittenberg, *Bautenschutz Bausanierung*, **13**, 98–101.

Owens, J.D. and Keddie, R.M. (1969) The nitrogen nutrition of soil and herbage coryneform bacteria, *Journal of Applied Bacteriology*, **32**, 338–45.

Pacovsky, R.S. (1990) Development and growth effects in the *Sorghum–Azospirillum* association, *Journal of Applied Bacteriology*, **68**, 555–63.

Poelt, J. (1991) Homologies and analogies in the evolution of lichens, *Frontiers in Mycology* (ed. D.L. Hawksworth), pp. 85–97, CAB International, Wallingford.

Read, D.J. (1991) Mycorrhizas in ecosystems – nature's response to the 'Law of the Minimum', *Frontiers in Mycology* (ed. D.L. Hawksworth) pp. 101–30, CAB International, Wallingford.

Richardson, D.H.L. (1991) Lichens and man, *Frontiers in Mycology* (ed. D.L. Hawksworth), pp. 187–210, CAB International, Wallingford.

Roberson, E.B. and Firestone, M.K. (1992) Relationship between desiccation and exopolysaccharide production in a soil *Pseudomonas* sp., *Applied and Environmental Microbiology*, **58**, 1284–91.

Sarkoh, N.A., Al-Hasan, R.H., Khaferer, M. and Radwan, S.S. (1995) Establishment of oil-degrading bacteria associated with cyanobacteria in oil-polluted soil, *Journal of Applied Bacteriology*, **78**, 194–9.

Schreiner, D.R. and Koide, R.T. (1993) Stimulation of a vesicular–arbuscular mycorrhizal fungi by mycotrophic and nonmycotrophic plant root systems, *Applied and Environmental Microbiology*, **59**, 2750–2.

Schuler, R. and Haselwandter, K. (1988) Hydroxamate siderophore production by ericoid mycorrhizal fungi, *Journal of Plant Nutrition*, **11**, 907–13.

Smith, R.I.L. (1990) Signy Island as a paradigm of biological and environmental change in antarctic terrestrial ecosystems, *Antarctic Ecosystems – Ecological Change and Conservation* (eds. K.R. Kerry and G. Hempel), pp. 32–50, Springer-Verlag, Berlin.

Truper H.G. and Galinski, E.A. (1986) Concentrated brines as habitats for micro-organisms, *Experientia*, **42**, 1182–7.

Tsuji, T., Kawasaki, Y., Takeshima, S. *et al.* (1995) A new fluorescence technique for visualizing living microorganisms in soil, *Applied and Environmental Microbiology*, **61**, 3415–21.

Vazquez, F.J., Acea, M.J. and Carballas, T. (1993) Soil microbial populations after wildfire, *FEMS Microbiology Ecology*, **13**, 93–104.

Wachtershauler, G. (1992) Groundworks for an evolutionary biochemistry: the iron–sulphur world, *Progress in Biophysics and Molecular Biology*, **58**, 85–201.

Xu, L.-H., Li, Q.-R. and Cheng, L.-J. (1996) Diversity of soil actinomycetes in Yunnan, China, *Applied and Environmental Microbiology*, **62**, 244–8.

Yang, C.-H., Menge, J.A. and Cooksey, D.A. (1994) Mutations affecting hyphal colonisation and pyoverdine production in pseudomonads antagonistic towards *Phytophthora parasitica*, *Applied and Environmental Microbiology*, **60**, 473–81.

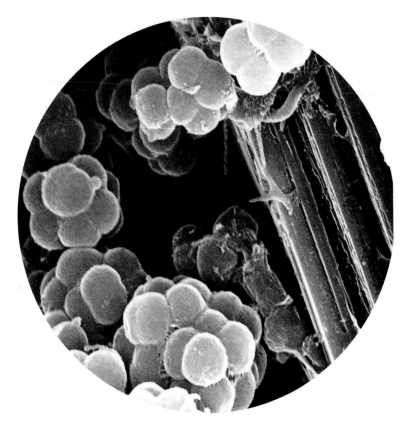

4 EXTREME ENVIRONMENTS

In the context of higher animal life many of the environments favoured by micro-organisms appear remarkably inhospitable. An environment, however, that permits the development of a wide diversity of micro-organisms, or that places only moderate selective pressure on its inhabitants is not regarded as extreme. Very low temperature environments, for example, are a common feature of life on this planet and are colonized by a wide range of organisms. In this chapter, four types of environment are considered extreme: very high temperature, highly acidic, highly alkaline and highly saline. There is considerable overlap in that many high temperature environments are also highly acidic and many highly alkaline environments are also highly saline. In each case conditions are unfavourable to the great majority of micro-organisms and a highly adapted and characteristic microflora develops.

Many of the micro-organisms developing under the most extreme conditions are archaebacteria (archaea). This has been attributed to conditions in extreme environments generally being similar to those that prevailed on earth in the earliest (archaeic) times.

HIGH TEMPERATURE ENVIRONMENTS

Occurrence and distribution of high temperature environments

High temperature environments are widely distributed and of many different types. The best known are probably hot springs, geysers and surrounding sulphur-containing soils (solfataras). Equivalent hydrothermal environments, such as vents and seamounts (underwater volcanoes), have also received considerable recent attention. Some oil fields, such as those of the North Sea are also high temperature environments, having much in common with hydrothermal systems. All of these are physically large environments, but localized high temperature environments occur world-wide in such places as spontaneously heated coal tips and compost heaps. In these situations, as well as in metal leaching operations (see page 139), steep temperature gradients may be present leading to thermophilic micro-organisms and their non-thermophilic counterparts growing in close proximity.

The works of man also create a large number of high temperature environments, for example heating systems, cooling operations in thermal power stations and chemical plant.

Thermophilic micro-organisms

Thermophilic representatives of algae, fungi, protozoa, cyanobacteria, eubacteria and archaebacteria are known. Thermotolerant bacteriophages have also been detected. Representative genera are listed in *Table 7*. Definitions of thermophilic micro-organisms vary somewhat but the term is generally accepted as being micro-organisms having an optimum growth temperature (T_{opt}) in excess of 45°C (115°F). Some care is required since the term can be used very loosely indeed.

Although thermophilic representatives of all types of micro-organisms exist, it is notable that eukaryotic thermophiles are restricted to an upper temperature of 61°C (141.8°F) while cyanobacteria isolated from alkaline hot springs and fissures of

Table 7. Representative thermophilic micro-organisms and their maximum temperatures for growth (T_{max}).

Thermophilic micro-organism	T_{max} °C (°F)
Algae	
Cyanidium caldarium	56 (133)
Fungi	
Talaromyces thermophilus	57–60 (134–140)
Chaetomium thermophile	58–60 (136–140)
Aspergillus candidus	50–55 (122–131)
Paecilomyces candidus	55–60 (131–140)
Thermomyces ibadensis	60–61 (140–142)
Dactylaria gallopava	50 (122)
Protozoa	
Cercosulcofer hemathensis	56 (133)
Vahlkampfia reichi	55 (131)
Cyanobacteria	
Synechococcus lividus	74 (165)
Oscillatoria amphibia	57 (135)
Phormidium laminosum	57–60 (135–140)
Mastigocladus laminosus	63–64 (145–147)
Eubacteria	
Bacillus stearothermophilus	70–75 (158–167)
B. coagulans	55–60 (131–140)
B. acidocaldarius	70 (158)
B. caldotenax	85 (185)
B. caldolyticus	82 (180)
Clostridium thermocellum	70 (158)
Cl. thermosaccharolyticum	67 (153)
Thermomicrobium roseum	85 (185)
Thermoactinomyces vulgaris	70 (158)
Desulfovibrio thermophilus	85 (185)
Thermus thermophilis	85 (185)
Thermotoga maritima	90 (194)
Archaebacteria	
Sulfolobus acidocaldarius	85 (185)
Acidianus infernus	95 (203)
Pyrobaculum islandicam	103 (217)
Pyrodictium occultum	110 (230)
Thermococcus celer	97 (207)
Pyrococcus furiosus	103 (217)
Methanothermus sociabilis	97 (207)
Methanococcus jannaschii	86 (187)
Thermoplasma volcanium	67 (153)

desert rocks will not grow above 74°C (165.2°F). Environments with temperatures above 74°C are exclusively the realm of eubacteria and archaebacteria. Thermophilic eubacteria may be placed in three classes: facultative thermophiles – these have a maximum growth temperature (T_{max}) of 50–65°C (122–149°F) but also grow at temperatures as low as 30°C (86°F). This category includes bacteria involved in the spoilage of canned foods, bacteria such as *Bacillus coagulans* and some strains of *B. stearothermophilus*.

Other strains of *Bacillus stearothermophilus* are obligate thermophiles meaning they have a minimum temperature (T_{min}) of *c*.40°C (104°F) and a T_{opt} of 65–70°C (149–158°F).

Extreme thermophiles (caldoactive bacteria) are eubacteria having $T_{min} > 40$°C, $T_{opt} > 65$°C and $T_{max} > 70$°C. These include Gram-positive, endospore-forming bacteria such as *B. caldolyticus* as well as Gram-negative genera such as *Thermus*. Care must be taken in using these definitions and a preferred approach may be to simply use the term thermophile with the proviso that the specific temperature growth range is given (Brock, 1984).

A further category of thermophiles also exists, the hyperthermophiles, having T_{min} of *c*.55°C (131°F) and $T_{max} > 80$–110°C (176–230°F). To some extent this category is clearly defined but the term should still be used with caution. A few hyperthermophiles are eubacteria, but the majority are archaebacteria. Although 110°C is currently the highest T_{max} known, this temperature has no particular biological or physical significance and it has been predicted, on the basis of the stability of macromolecules, that bacteria similar to known species could grow at temperatures as high as 150°C (302°F).

It is likely that a number of superthermophiles exist in hydrothermal environments where high pressures maintain water in the liquid form. Evidence for this has been obtained by finding significant levels of particulate DNA in superheated smoker fluids at temperatures of 174–357°C (345–675°F) (Straub *et al.*, 1990). It has also been suggested that bacteria are able to thrive in the unexplored habitats considerably below accessible vent formations. This suggestion is based on the observation that a sulphate-reducing, heterotrophic hyperthermophile (T_{opt} 100°C (212°F)) had a trend toward barophily at pressures considerably in excess of those encountered on the ocean floor (Reysenbach and Deming, 1991).

Some thermophilic eubacteria, such as *Thermotoga maritima*, have highly unusual properties. Studies of the *T. maritima* 16S rRNA have shown this organism to represent the deepest known, branching in the eubacterial line of descent and also the most slowly evolving of eubacterial lineages. Slowly evolving lineages tend to retain ancestral characteristics and the earliest eubacteria may have been thermophiles (Friedman, 1992).

The archaebacterial domain contains two main branches, one lineage contains the methanogens, some of which are hyperthermophiles, while the other contains the sulphur-metabolizing hyperthermophiles. The hyperthermophile *Archaeoglobus* appears to form a separate branch, intermediate between methanogens and sulphur-metabolizers. Hyperthermophiles may be heterotrophs or obligate/facultative autotrophs. All, with the exception of *Acidianus*, *Metallosphaera* and *Sulfolobus*, are strict anaerobes. The aerobic bacterium *Thermoplasma*, which is present in self-heating coal refuse, is not a true hyperthermophile (T_{max} 70°C) but is very closely related. Hyperthermophiles, such as *Pyrodictium occultum* (T_{min} 82°C (180°F)), are able to exploit high temperature eco-niches very effectively, but have a highly restricted biotope. In contrast, extremely thermophilic eubacteria can often grow over a temperature range as great as 40°C. Bacteria capable of growth over such a wide range, which include *B. stearothermophilus*, *Cl. thermohydrosulfuricum* and *Thermoanaerobacter ethanolicus*, exhibit biphasic Arrhenius plots of growth temperature vs. doubling time. This results from the temperature-regulated synthesis of two sets of key enzymes, which differ in temperature stability.

Some high temperature environments are also highly acidic and populated by a specialist group of thermophiles, the thermoacidophiles. These bacteria, including *Sulfolobus* and *Thermoplasma*, are able to grow at high temperatures at pH values of 1–2 (*Feature 14*, page 134).

Feature 14. An industrial revolution

There is considerable interest in the use of thermophilic micro-organisms in biotechnology. Fermentations proceed more rapidly at high temperatures and there is less concern over contamination. There has been particular interest in the use of thermophilic anaerobic fermentations in the production of fuel and chemicals (*Table 8*). Enzymes derived from thermophilic bacteria are also of interest due to their high level of heat stability. Considerable use is already made of thermostable enzymes, including a xylose isomerase from *B. coagulans* used in the manufacture of high-fructose corn syrup at 60°C (140°F). On a smaller scale, thermostable DNA polymerase is produced for use in the polymerase chain reaction; the use of heat-stable proteins from the same source has been proposed in the manufacture of semiconductor chips.

Microbial communities in thermophilic environments

Hydrothermal vents and seamounts

The hydrology of hydrothermal vents has been discussed in the context of the ocean as a whole in 'Deeps-sea habitats' (see pages 40–42). Thermophiles are important as both producers and consumers of organic matter. Among the more important genera are *Pyrodictium occultum*, *Pyrococcus* and *Thermococcus* as well as poorly defined archaebacterial isolates. *Archaeoglobus* appeared to be present only in a hydrothermal system off Italy, but has now also been isolated from a volcanic slick from an erupting Polynesian seamount and from water in a geothermally heated North Sea oil well (see below). An open-sea volcanic plume from the erupting MacDonald seamount contained at least 10^6 hyperthermophiles per litre, a number of which were closely related to existing archaebacteria including

Archaeoglobus (Haber *et al.*, 1990). This was considered to provide evidence of spreading and long-term survival of hyperthermophiles at temperatures below those supporting growth. Some isolates were, however, unique to the erupting seamount, including isolate MD7 (**280**), an obligate anaerobe recovered by incubation at 100°C (212°F) for 24 hours under a pressure of 300 kPa.

Although the seamount ecosystem was not understood, it was thought that the production of organic matter resulted from the activity of the anaerobic H_2-oxidizers *Pyrodictium* and *Archaeoglobus*. These micro-organisms, which use CO_2 as the sole carbon source, obtain energy by sulphur and SO_4^{2-} respiration respectively. The seamount community also contains strictly heterotrophic anaerobes including *Pyrococcus*, *Thermococcus* and MD7. These organisms are consumers of organic matter, growing by fermentation, or sulphur

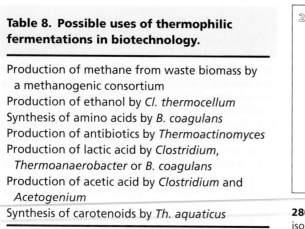

Table 8. Possible uses of thermophilic fermentations in biotechnology.

Production of methane from waste biomass by a methanogenic consortium
Production of ethanol by *Cl. thermocellum*
Synthesis of amino acids by *B. coagulans*
Production of antibiotics by *Thermoactinomyces*
Production of lactic acid by *Clostridium*, *Thermoanaerobacter* or *B. coagulans*
Production of acetic acid by *Clostridium* and *Acetogenium*
Synthesis of carotenoids by *Th. aquaticus*

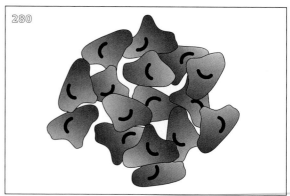

280 Diagrammatic representation of hyperthermophilic isolate MD7 from an erupting seamount.

respiration, on proteins, sugars and the like. (Haber *et al.*, 1990). Synthesis of amylolytic enzymes in response to complex carbohydrates has been reported in *Pyrococcus*; polysaccharide availability appears to be an important aspect of growth in hydrothermal environments.

Hot springs

Hot springs and nearby soils vary both in temperature and pH value, with obvious consequences for the microbial communities developing. Light intensity is also of importance where photosynthetic micro-organisms are present. In non-acid springs, at temperatures below 74°C (165°F), cyanobacteria are of major importance in primary production and, in the absence of grazing by predators, microbial mats develop. Like microbial mats in other environments, these exhibit both vertical and temporal heterogeneity. Cyanobacteria and green non-sulphur bacteria dominate the upper 1 mm but are not usually present at depths below 2.5 mm, probably as a result of phototrophic behaviour. *Synechococcus* is the most common culturable cyanobacterium while the dominant culturable green non-sulphur bacterium is the filamentous gliding bacterium *Chloroflexus*. *Synechococcus* and *Chloroflexus* grow in close association, although *Synechococcus* is favoured by mildly acid conditions and *Chloroflexus* by mildly alkaline conditions and higher temperatures. During growth in the light, *Synechococcus* accumulates carbohydrates, which are then used during growth in the dark. Some 10% of photosynthetically fixed carbon is excreted by *Synechococcus* as organic carbon, which is available to *Chloroflexus*. Some of the metabolic transformations which occur in the light may be due to *Chloroflexus* and a low level of heterotrophic uptake is possible in the dark. Irrespective of the mechanism, accumulation of C-storage molecules in the light can be used to drive the synthesis of protein in the dark. Photosynthetic activity by *Chloroflexus* is limited by the presence of oxygen in upper layers of the mat and consequent repression of bacteriochlorophyll synthesis.

Anaerobic, thermophilic, glycolytic bacteria have been isolated from various hot springs. The major nutrient source appears to be xylans, derived from vegetation blown into the spring. Most of the xylan-utilizing bacteria are Gram-negative, asporogenous, non-motile rods, identified with *Dictyoglomus*.

The situation in hot, acid springs is different, in that cyanobacteria and green non-sulphur bacteria

281 A diagrammatic representation of *Sulfolobus*.

are unable to grow and mats do not develop. Environments of this type are dominated by hyper-thermophiles, especially *Sulfolobus* (**281**). This organism has an optimum pH value of *c*.2.0 although growth occurs over the range from pH 1.0 to 5.9. In hot springs and their immediate surroundings, *Sulfolobus* grows as a respiratory chemoautotroph. The principle substrate for growth is elemental sulphur, the bacterium growing on the surface of crystalline sulphur and oxidizing it to sulphuric acid. Elemental sulphur is formed in hot springs by the leaching of iron and sulphide by geothermal steam or hot water. Sulphide is then oxidized to elemental sulphur by either chemical or biological processes involving oxygen or ferric iron. *Sulfolobus* is very widely distributed in hot springs, but other thermoacidophiles appear to have a limited distribution. Anaerobic archaebacteria of the *Thermoproteus* group, for example, have been isolated only from geothermal regions of Iceland. This group originally comprised two genera (the rod-shaped *Thermoproteus* and the coccal *Desulfuro-coccus*) and has a minimum pH value for growth of 5.0 and is not, therefore, found in strongly acid springs. In nature, the organisms, despite being strict anaerobes, have a predominantly respiratory metabolism with elemental sulphur serving as the terminal electron acceptor. More recently, a related bacterium, *Pyrobaculum aerophilum*, has been described which is capable of aerobic growth or nitrate respiration.

Geothermally heated oil wells

These have much in common with hydrothermal systems in terms of their physics and chemistry. Wells are typically drilled to a depth of 2–4 km where temperatures are in the range 60–130°C (140–266°F) and pressures between 20–50 mPa. Most wells contain formation water (*in situ* pore water) that has a low SO_4^{2-} content and does not support significant growth of sulphate-reducing bacteria. It is common practice, however, to enhance oil recovery by flooding wells with anoxic seawater, which may stimulate thermophilic, sulphate-reducing bacteria capable of growth on organic acids. The resultant H_2S production can cause severe problems, through corrosion of metal alloys used in wells and processing equipment. Precipitation of sulphides may cause extraction problems while H_2S itself reduces oil quality by souring and presents a health hazard to personnel working on production platforms.

Thermophilic strains of *Desulfotomaculum* are widespread and *Thermodesulfurobacterium* are also common. A new species of *Archaeoglobus*, *Arch. fulgidus*, has been isolated from high temperature wells where it probably grows in the well-head and oil–water separator at temperatures between 70 and 85°C (158 and 185°F). In moderately thermophilic wells (65–75°C (149–167°F)) all three species are able to coexist, possibly growing in association with fermentative bacteria producing H_2 and valerate for mixotrophic growth. Although the sulphate content is low, sulphate accumulation and high-affinity uptake permits growth.

Self-heating coal refuse

Coal refuse containing significant quantities of residual coal and iron pyrites (FeS) is prone to heating due to the oxidation of FeS by chemo-autotrophic bacteria. This also results in acidification, creating suitable conditions for the growth of the enigmatic bacterium *Thermoplasma*. This organism is found only in coal refuse with a temperature between 37 and 65°C (98.6 and 149°F) and a pH value between 0.5 and 4.5. *Thermoplasma* has not been isolated from any naturally occurring hot environment and it has been speculated that the natural habitat is coal veins. Coal, pyrrolysed by heating, supplies *Thermoplasma* with growth factors. The nature of these is not known but the requirement can be satisfied by an oligopeptide from yeast extract.

Charcoal piles

Charcoal piles provide an eco-niche to a highly specialized and adapted thermophilic actinomycete, *Streptomyces thermoautotrophicus*, a CO^- and H_2^- oxidizing bacterium. Charcoal piles consist of burning wood covered with a layer of soil. The temperature profile of the soil closely matches that of the organisms' growth while smoke provides a constant source of CO and H_2. Condensing water vapour maintains the soil in a moist condition and minerals are obtained from ashes. Charcoal burning is an intermittent process but *Str. thermoautotrophicus* has at least two mechanisms permitting extended survival. Exospores are produced, which withstand desiccation and temperatures up to 100°C (212°F). At the same time, oxidation of CO at temperatures down to 10°C (50°F) allows the organism to meet its maintenance requirements although not to grow (Gadkari *et al.*, 1990).

EXTREMELY ACIDIC ENVIRONMENTS

Occurrence and distribution of extremely acidic environments

Although many environments, including some foods, have pH values significantly below neutrality, these are not regarded as extremely acidic. Extremely acidic environments (pH value $c.3.0$ or less) are not widespread in nature – the most common are strip coal mining operations, coal waste heaps, metal ore spoil heaps and drainage water from ore mines. These environments are typically high in sulphate, pyrite and H^+ ions. Acidity is derived from the oxidation of sulphides and pyritic material to sulphuric acid. Above pH 4.5 abiotic oxidation occurs at a slow rate. Below this pH oxidation is mediated by bacteria and proceeds at a rate some 10^6 times faster. Some acid peat has a pH value of $c.2.0$ and is also considered extremely acidic. As discussed above (see page 133) some hot springs and surrounding acid soils are also extremely acidic and populated by thermoacidophiles. Extremely acidic environments are found in association with chemical processing including laboratory reagents. In general, however, no attempt has been made to study the micro-organisms present.

Acidophilic and acid-tolerant micro-organisms

Micro-organisms able to grow in extremely acidic environments comprise both acidophiles and acid-tolerant types. Various definitions of acidophile exist, but in the current context these are considered to be micro-organisms that grow optimally at a pH of 3.0 or less but not close to neutrality. The majority of species of acidophiles are thermo-acidophilic archaebacteria and so are discussed in the context of high temperature environments. Some of these micro-organisms, however, are of very restricted distribution. Non-thermophilic acidophiles are eubacteria which, although belonging to a smaller number of genera, are much more widespread and more versatile. The best known genus is *Thiobacillus* which has a pH_{opt} of $c.2.0$. The two most common species are *Th. thiooxidans* (which utilizes elemental sulphur as its energy source) and *Th. ferrooxidans* (which can utilize both reduced sulphur compounds and Fe^{2+}). *Th. ferrooxidans* is of considerable interest in the recovery of metals from ores by bioleaching (see below). Moderate thermophiles resembling *Th. ferrooxidans* have also been isolated, while a second Fe^{2+}-oxidizing thiobacillus, *Th. prosperus*, is a moderate halophile. In the past *Thiobacillus* and other members of the genus have often been considered to be obligate chemolithotrophs but a mixotrophic strain of *Th. ferrooxidans* has been described and *Th. acidophilus* and *Th. cubrinus* both have a mixotrophic metabolism, obtaining energy by the oxidation of reduced sulphur compounds but not Fe^{2+}.

Leptospirillum ferrooxidans is a second species of iron-oxidizing acidophile which, in contrast to *Th. ferrooxidans*, is incapable of oxidizing reduced sulphur compounds. Some strains are moderately thermophilic. Moderately thermophilic, iron-oxidizing, endospore-forming bacteria have also been isolated. In contrast to *Leptospirillum* and *Thiobacillus*, these bacteria can grow as chemolithoheterotrophs, obtaining energy from Fe^{2+} oxidation and carbon from organic substrates.

Heterotrophic acidophiles were first isolated fairly recently. The best characterized genus is *Acidophilium*; this grows in close association with *Thiobacillus* and is a common contaminant in cultures of thiobacilli. *Acidophilium* shares a number of properties with *Thiobacillus*, although there are significant differences in carbon and energy metabolism. Heterotrophic acidophiles have also been isolated which resemble neutrophilic bacteria of the *Sphaerotilus–Leptothrix* group in morphological and behavioural characteristics but which grow in the pH range 2.0–4.4.

Although most eukaryotic micro-organisms in low pH environments are acid-tolerant rather than acidophilic, the algae *Chlamydomonas acidophila* and *Euglena mutabilis* reportedly each have a pH_{opt} of $c.3.0$ and a pH_{max} of $c.5.5$.

Acid-tolerant organisms are those able to grow

at low pH values but which have a significantly higher pH_{opt}. Relatively few non-acidophilic bacteria are capable of good growth at pH values below 4.0 and, with important exceptions such as *Metallogonium*, are not found in the same habitats as acidophiles. Yeasts are able to grow at pH values as low as 2.0 while fungi have the widest pH growth range, strains of *Aspergillus*, *Fusarium* and

Penicillium reportedly growing over the range pH 2–10. Acid-tolerant algae are known, some species of *Chlorella* are able to grow at pH 2.0 but have a pH_{opt} of *c.*7.0. Flagellate protozoa, resembling *Eutrepia* and several amoeboid protozoa have been isolated from water at pH 2.0 and while actual growth was not demonstrated for a number of years, grazing has now been observed.

Microbial communities in extremely acidic environments

Ore spoil heaps and mine drainage water

Development of acidity in ore spoil heaps initially involves the slow abiotic formation of sulphuric acid followed by a microbial succession in which *Metallogonium* lowers the pH value sufficiently for growth of *Thiobacillus*.

Although *Thiobacillus* is numerically dominant it is found in close association with *Acidophilium* and, where sufficient Fe^{2+} is available, *Leptospirillum*. There is evidence of adaptation to the ambient temperatures at the mine site, in that psychrotrophic strains of *Thiobacillus* and *Acidophilium* have been isolated from a Canadian uranium mine where temperatures do not usually exceed 18°C (64.4°F). At the same time moderately thermophilic strains of *Thiobacillus* and *Leptospirillum* have been isolated from heated spoil heaps. Such isolates, however, are able to grow over a wide temperature range, including higher ambient temperatures. In addition to acido-

philic bacteria, yeasts resembling *Rhodo-torula* and various protozoa have been observed in acidic mine drainage water although numbers are generally small.

Heterotrophic acidophiles, apparently related to the *Sphaerotilus–Leptothrix* group, were isolated from a disused pyrites mine in Wales. The mine was characterized by the copious growth of gelatinous 'acid streamers' in the main drainage stream and on moist surfaces. These consisted of a mixed community of filamentous and unicellular bacteria, the latter probably being *Th. ferrooxidans*, *Ls. ferrooxidans* and *Acidophilium*. The bacteria were being grazed by protozoa and rotifera (*Feature 15*).

Micro-organisms associated with ores such as copper require a high tolerance of the metal ions. Metal resistance does not appear to be an inherent property of the micro-organisms, but results from reduced interaction with the metals. The mechanism involves competition for metal-sensitive sites between toxic metal cations and the excess H^+ ions usually present (Brierley *et al.*, 1980).

Feature 15. The banks of the Ohio

The large quantities of sulphuric acid produced by *Thiobacillus* inevitably enter drainage water and are a serious source of pollution. During the 1960s it was estimated that 3×10^4 tons of sulphuric acid entered the Ohio River annually as a result of pollution from mine drainage. This is probably an extreme example but localized acid pollution undoubtedly occurs elsewhere. In Wales, for example, acid pollution by water emerging from the addit of a long-disused pyrites mine can be seen to affect the immediate environment, the stream containing massive growths of a *Gallionella*-like organism. *Gallionella* is not, however, usually associated with acidic environments (see also above).

Ore bioleaching operations

Bioleaching involves the use of micro-organisms to solubilize the metals contained in mineral ores. This enables the extraction of the mineral as an aqueous solution. Bioleaching is environmentally attractive since there is no air pollution, although engineering precautions must be taken to prevent contamination of groundwater. The process makes possible the recovery of metal from low-grade (lean) ores, metal that cannot be recovered by conventional pyrometallurgical processes because of an unfavourable ratio of host rock to ore mineral. The process is also economically attractive because it is neither energy nor labour intensive. Bioleaching is currently used for the solubilization of sulphidic copper and uranium ores, although extension to other ore types and other metals is perfectly feasible.

Thiobacillus ferrooxidans is the most important organism in bioleaching although *Th. thiooxidans* and *Ls. ferrooxidans* are also involved. *Acidophilium* is present in leaching operations and is thought to enhance activity of *Th. ferrooxidans* by removing inhibitory organic compounds. *Acidophilium* may also play a direct role in leaching through its apparent ability to oxidize elemental sulphur in the presence of an organic energy source. *Acidophilium* is also able to reduce Fe^{3+} to Fe^{2+} and thus regenerate an energy source for *Th. ferrooxidans* while contributing to the leaching process. Other micro-organisms, including fungi and protozoa, are present in the community which develops but relationships and consequences for metal recovery are little understood.

Ore leaching processes are of two types: *in situ* leaching of the ore in place and ore leaching after recovery by conventional mining. In each case a 'barren' solution containing ferric sulphate and sulphuric acid is allowed to percolate through the ore body. This creates conditions which enrich for an acidophilic microflora. In some cases, however, the process is accelerated either by recycling barren fluid containing acidophilic micro-organisms or by inoculating the barren fluid with desired micro-organisms. Conditions in ore leaching environments are highly favourable for growth of acidophiles although a particularly high level of metal tolerance is required. Under some circumstances there is a considerable degree of heating, favouring the growth of moderately thermophilic strains and, in localized areas, temperatures may be sufficiently high to support growth of SO- and Fe^{2+}-oxidizing strains of *Sulfolobus*. A further use of acidophiles in metallurgy is the biobeneficiation of gold ores. This involves use of *Th. ferrooxidans* to remove pyrites and arsenopyrites from sulphidic gold ores. This prevents interference from the pyrites during chemical extraction of the gold.

Strip coal mining and coal waste heaps

As an environment for acidophiles, pyrites-containing coal and coal waste resembles ore spoil heaps and the microflora is generally similar. Acid-tolerant fungi are more prevalent and can be isolated both from coal waste and drainage water at pH 2–3. A considerable degree of heating may occur, creating suitable conditions for *Thermoplasma* (*Feature 16*).

Feature 16. Down the pit

Acidophilic bacteria cause problems in strip mining operations, causing a loss of coal quality. Conversely, acidophiles have been proposed as a means of desulphurizing coal and thus reducing SO_2 emissions and acid rain. At present, however, the large-scale exploitation of acidophiles in this way does not seem practical due to the recalcitrant nature of organic sulphur compounds. It is also argued that desulphurization cannot be justified economically and that large scale electricity generation, a major source of SO_2, should be progressively switched to alternative energy sources. These include nuclear power, use of other fossil fuels (such as natural gas), wind and tidal power.

HIGHLY ALKALINE ENVIRONMENTS

Occurrence and distribution of highly alkaline environments

Stable alkaline environments (pH > 10) arise where a combination of ecological, geographical and climatic conditions occur, which mitigate against the buffering effect of atmospheric CO_2. Soda lakes and deserts are found in many parts of the world including North, Central and South America, Australia and in many parts of Africa. In Asia soda lakes occur in Turkey, India and China and in Europe in Hungary, Yugoslavia and the former USSR. Springs, of high Ca^{2+} content are a further type of naturally occurring, highly alkaline environment found in a number of geographical locations, including California, Oman, Cyprus, the former Yugoslavia and Jordan.

A number of unstable, man-made, alkaline environments are associated with use of alkalis (usually NaOH) as part of an industrial process or are derived from the generation of $Ca(OH)_2$. Alkali is used in a wide range of industrial processes including electroplating, food manufacture and paper making, while $Ca(OH)_2$ is generated during cement manufacture and in blast furnace waste. Also, transient, highly localized, alkaline niches arise from biological activity in many types of environment, such as soils, which are classified on the basis of overall pH as neutral or even acidic.

Alkaliphilic and alkalitolerant micro-organisms

A number of definitions may be applied to micro-organisms from highly alkaline environments. Alkaliphiles may be defined as having a pH_{opt} for growth of more than 8.0. A more common pH_{opt} is from 9.0–10.0 while some species grow at pH values as high as 11.5 (*Feature 17*). A number of micro-organisms are able to grow at relatively high pH values but have a pH_{opt} of 7.0 or below. These micro-organisms are referred to as alkali-trophic or alkalitolerant. Alkaliphiles that are unable to grow at pH 7.0 are classified as obligate alkaliphiles while a further subdivision is made into halophilic alkaliphiles (haloalkaliphiles), also able to grow at NaCl concentrations up to saturation, and nonhalophilic alkaliphiles. Haloalkaliphiles are archaebacteria and only two genera are currently recognized, *Natronobacter* and *Natronococcus*. Other micro-organisms capable of growth at high pH values, however, are of much greater diversity.

Feature 17. Washday blues

Alkaliphiles are of commercial interest in the production of enzymes, such as alkaline protease, which are stable at high pH values. Alkaline proteases are widely used in biological washing powders. Industrial use is also made of a cyclodextrin glucanotransferase in the manufacture of cyclodextrin. This enzyme has activity over the pH range 4.5–10 and a conversion rate in excess of 75–80%, compared with 50% for the *B. macerans* enzyme conventionally used. This permits crystallization of the dextrin directly from the primary starch–enzyme reaction mixture. It should be appreciated, however, that alkaline enzymes are present in some neutrophilic bacteria, *B. licheniformis*, for example, being a source of some alkaline proteases used in washing powders.

Microbial communities in highly alkaline environments

Soda lakes and deserts

Soda lakes and deserts have a pH in the range 10.5–11.0. At these pH values, Ca^{2+} and Mg^{2+} precipitate as carbonates and so only very low levels are present in soluble form. Sodium carbonate is usually present as $Na_2CO_3\,10H_2O$ or $Na_2CO_3\cdot Na_2CO_3\,2H_2O$ and there is a high, although variable, NaCl content. The best characterized soda lakes are those of the East African Rift Valley where both limnological and microbiological assays have been carried out for a number of years. The genesis of alkalinity in the Rift Valley has been attributed to vulcanism and soda-rich larva flows. Alternatively, alkalinity has been associated with sulphate reduction in anaerobic basins. These situations may not, however, be typical and, in general, soda lakes appear to arise from a particular combination of geological and climatic features (Grant, 1992). Low levels of Ca^{2+} and Mg^{2+} in the surrounding topography appear to be particularly important, together with the existence of distinct, isolated drainage basins and high rates of evaporation. These conditions are met in the Rift Valley which has a high Na^+, low Ca^{2+} and Mg^{2+} geology. Evaporative concentration has been of importance in the formation of the large number of lakes on the valley floor. These have a pH of 11.0–11.5 and vary in salinity from 5% in the north to saturation in lakes such as Lake Natron in the south. Soda lakes have a very high buffering capacity which maintains a constant high pH in conditions of varying rainfall. In some soda lakes, Ca^{2+} groundwater, which may have travelled a considerable distance, results in localized calcite ($CaCO_3$) precipitation and formation of 'tufa' columns. Cyanobacteria are involved in formation of such structures through the biogenic precipitation of aragonite, another crystalline form of $CaCO_3$.

Soda lakes have been described as the most productive naturally occurring aquatic environments, primary production being in excess of an order of magnitude greater than that of non-alkaline streams and lakes (Grant, 1992). To some extent this is due to the location of soda lakes in areas of high ambient temperature (30–35°C (86–95°F)) and a high intensity of illumination rather than to the specific properties of these environments. A specific factor in the very high productivity, however, is the virtually unlimited supply of HCO_3^- for photosynthesis. Prokaryotes are almost entirely responsible for primary production in soda lakes and blooms of phototrophs are common. To a large extent the NaCl content determines phototroph growth and more dilute lakes, with salinities of 5–15% NaCl, often exhibit permanent or seasonal cyanobacterial populations of only one or two genera at high cell densities. The dominant cyanobacteria are often poorly characterized but include the filamentous *Cyanospira* and the unicellular *Chroococcus* and *Pleurocapsa*. In some cases, blooms consist almost entirely of filamentous cyanobacteria referred to either as *Spirulina* (*Feature 18*) or *Arthrospira*. These cyanobacteria are gas vacuolated and float as a scum on the surface of less saline lakes. In some lakes *Spirulina* supports enormous populations of flamingos, Lake Nakura (Rift Valley) having an

---

Feature 18. Our daily bread

Spirulina is the only cyanobacterium consumed as human food, having been eaten by the Aztecs of Mexico and consumed today in Chad. In the western world, *Spirulina* is considered a health food due to its favourable amino acid content and low level of nucleic acids. In the sporting world, *Spirulina* had a vogue as a legal means of enhancing performance. This raises the question of whether it should be more permissive to use *Spirulina* than, for example, a steroid to improve performance. *Spirulina* is also of interest in biotechnology, its very high photosynthetic yield resulting in an efficient means of producing biochemicals such as essential amino acids and water-soluble vitamins.

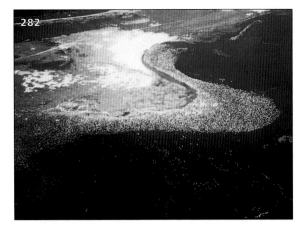

282 Flamingos on Lake Nakura: a massive population supported by *Spirulina*.

283 A pink surface crust formed by *Natronobacterium* on the surface of Lake Magadi. **282** and **283** reproduced by courtesy of Prof. W.D. Grant, Department of Microbiology and Immunology, University of Leicester, UK.

estimated population of 2×10^6 birds consuming 200 tons of cyanobacterial biomass each day (**282**).

Eubacteria also play a role in primary production and the contribution of the anoxygenic phototroph *Ectothiorhodospira* is significant during this period; this bacterium blooms alongside cyanobacteria. The bacterium is also involved in the sulphur cycle through utilization of H_2S as electron donor in photosynthesis.

Less saline lakes also contain non-phototrophic bacteria at population densities of 10^7–10^8 ml^{-1}. Populations are stable and unaffected by cyanobacterial blooms or dilution after rainfall. The population is largely Gram-negative, aerobic bacteria, which cannot be assigned to existing taxa. Anaerobic non-phototrophs are also present and the presence of black anoxic sediments provides circumstantial evidence for the presence of alkaliphilic, sulphate-reducing bacteria. Methanogenesis is also a feature and is attributed to the methylamine-utilizing methanogen *Methanohalophilus*.

Highly saline soda lakes, which have an NaCl content ranging from 20% to saturation, have a different microflora. Cyanobacteria are absent and the dominant micro-organisms are haloalkaliphilic archaebacteria. These reach populations of 10^7–10^8 ml^{-1}. Haloalkaliphiles are classified in two genera: *Natronobacter* (**283**), which contains three species, and *Natronococcus*, which has a single species *Na. occultus*. Archaebacteria of this type occupy a highly specialized eco-niche, which does not overlap with that occupied by other halophilic archaebacteria. The ecology of haloalkaliphiles has been considered analogous to that of non-alkaliphilic halophilic archaebacteria in neutral salterns. Alkaline salterns, such as Lake Magadi (Rift Valley) and Wadi Natrum (Egypt), undergo a cycle of precipitation of both NaCl and Na_2CO_3, however, which is totally unknown in neutral salterns (Grant, 1992).

Haloalkaliphilic archaebacteria are secondary producers in highly saline soda lakes and, in the absence of cyanobacteria, halotolerant strains of *Ectothiorhodospira* are responsible for primary production. Other halotolerant, alkaliphilic eubacteria are present which, although poorly characterized, appear to represent new taxa. These organisms are present only in small numbers in lakes approaching saturation but can compete effectively with haloalkaliphiles at NaCl concentrations of 15–20% when numbers of the two types are approximately equal. Methanogenesis due to *Methanohalophilus* continues as the NaCl concentration approaches saturation and the presence of sulphate-reducing bacteria is strongly predicted but as yet unproven.

High Ca^{2+} alkaline springs

Water of this type is chemically similar to that produced during cement manufacture. Alkalinity is generated in the virtual absence of CO_3^{2-} ions which are removed from solution in the presence of excess Ca^{2+} and Mg^{2+}. In the initial stages, CO_2-charged surface water erodes surrounding minerals. The chemistry of the groundwater forming the spring is determined by low temperature, sub-surface weathering of two primary minerals, olivine and pyroxene, which are decomposed by the process of serpentization (**284**). This results in a $Ca(OH)_2$ brine in which the $Ca(OH)_2$ is in equilibrium with the solid phase. Under most circumstances, the solubility of $Ca(OH)_2$ is very low and the brine is very dilute, being effectively an oligotrophic system in comparison with the eutrophic soda lakes. The buffering capacity is very weak when separated from the solid phase $Ca(OH)_2$ and exposed to atmospheric CO_2. Magnesium ions are removed during production of serpentine, while CO_3^{2-} is rapidly removed from solution as carbonates. Conditions are predominantly reducing due to the release of Fe^{2+} ions and the hydrolysis of transient metal hydroxides with the production of H_2.

Relatively few microbiological studies have been made of high Ca^{2+} alkaline springs. Eubacteria may be isolated at levels of up to 10^4 ml^{-1}. Isolates, however, are not particularly alkaliphilic and are bacteria common in other environments. It appears likely that these bacteria are contaminants rather than members of an alkaliphilic microflora. The question of whether or not micro-organisms can grow in these environments remains unresolved.

284

$$MgFeSiO_4 + CO_2 + H_2O \rightarrow$$
olivine

$$Mg^{2+} + HCO_3^- + H_4SiO_4 + Fe^{2+} + OH^-$$

$$MgCaFeSiO_3 + CO_2 + H_2O \rightarrow$$
pyroxene

$$Mg^{2+} + Ca^{2+} + HCO_3^- + H_4SiO_4 + Fe^{2+} + OH^-$$

$$Mg^{2+} + H_4SiO_4 + H_2O \rightarrow Mg_3Si_2O_5(OH)_4 + H^+$$
$$\text{serpentine}$$

$$MgFeSiO_4 + MgCaFeSiO_3 + H_2O \rightarrow$$
olivine   pyroxene

$$Mg_3Si_2O_5(OH)_4 + Fe^{2+} + Ca^{2+} + OH^-$$
serpentine

$$Fe(OH)_2 \rightarrow Fe_2O_3 + H_2 + H_2O$$

Note: Based on Barth *et al.* (1987)

284 Geochemical evolution of high Ca^{2+} alkaline springs.

Alkaliphilic micro-organisms from non-alkaline environments

Alkaliphilic micro-organisms can be isolated fairly easily from neutral and even acid environments, especially soils, and to a lesser extent water. It is postulated that alkaliphiles are able to alter the pH value to permit growth in a very small niche. Alternatively the metabolic activity of other micro-organisms may raise the pH sufficiently for alkaliphiles to develop. The majority of alkaliphilic micro-organisms from non-alkaline environments are Gram-positive, endospore-forming rods, currently classified with *Bacillus*. Most isolates have a red or yellow pigmentation and grow over the pH range 8.0–11.5.

Non-endospore forming alkaliphiles from non-alkaline environments have been identified with a wide range of genera. These micro-organisms are present in soil and water, although at significantly lower levels than endospore-formers. In some cases the isolates are alkalitrophic rather than alkaliphilic and, in others, the term has been loosely applied. A few alkaliphilic yeasts and moulds have been isolated but the majority of non-endospore formers are bacteria. These include isolates identified with *Aeromonas*, *Corynebacterium*, *Micrococcus*, *Paracoccus*, *Pseudomonas* and *Streptomyces*.

Man-made alkaline environments are populated by alkaliphilic and alkalitrophic bacteria of the same types as are isolated from non-alkaline environments. It is assumed that these are derived by enrichment from nearby soil and water. In some cases a highly specific microflora develops. Examples are *Exiguobacterium aurantiacum* in potato processing waste and *Ancyclobacterium* in waste water from paper and board processing (Grant, 1992).

NON-ALKALINE, HIGHLY SALINE ENVIRONMENTS

Occurrence and distribution of non-alkaline, highly saline environments

Large-scale, non-alkaline, saline environments are typified by concentrated brines such as those in solar salterns and salt lakes where NaCl concentrations may be saturated. Many environments of this type are situated in tropical or subtropical regions but some are found in temperate regions; salt lakes also exist in Antarctica. Concentrated brine environments are of two basic types: thalassohaline (derived from seawater and characterized by a neutral to slightly alkaline pH value and a dominance by Na^+ and Cl^- ions) and athalassohaline (in which ions other than Na^+ and Cl^- are of major importance). Such environments include soda lakes.

Halophilic archaebacteria survive in both solar and mined salt; small-scale, highly saline environments also exist in fish salting and leather tanning operations as well as in brines used in meat curing. Hypersaline soils are found in the bottoms of dried-up salt lakes and abandoned salterns but are of relatively low NaCl content ($c.5.0\%$ to $c.11\%$). A novel terrestrial halophilic environment is represented by the plant leaf surfaces of a salt-excreting plant *Atriplex halimus*.

Halophilic micro-organisms

These may be defined as those with an NaCl optimum requirement for growth in excess of physiological levels: obligate halophiles are those unable to grow at normal physiological levels while facultative halophiles are those which are able to grow at normal physiological levels but which grow optimally at higher concentrations. Halotolerant organisms are able to grow at relatively high NaCl concentrations but grow optimally at normal physiological levels. It must be stressed that neither facultative nor obligate halophiles are necessarily able to grow in highly saline environments. Halophilic micro-organisms able to grow in highly saline environments of greater than 15–20% NaCl comprise both eubacteria and archaebacteria. Fungi and, allegedly, some yeasts are able to grow at NaCl concentrations approaching saturation but these micro-organisms are halotolerant rather than halophilic.

Although a number of halophilic eubacteria are able to grow in NaCl concentrations approaching saturation, the optimum is usually considerably lower. Some of these organisms grow over a very wide growth range and a requirement for NaCl at levels above physiological can be hard to demonstrate. Minimum NaCl concentrations are usually 2.5–4% and optimum concentrations typically 8–10% although some have an optimum as low as 5%. Bacteria of this type are commonly referred to as moderate halophiles. Some genera, such as *Halomonas*, contain no non-halophilic species, but many are members of genera common in normal environments. Gram-negative rods, such as *Vibrio* and *Pseudomonas*, are most common but a halophilic species of *Bacillus* is also recognized. A number of obligately anaerobic halophiles have been described, including *Halobacteroides* and *Sporohalobacter*. Halophilic members of specialist groups have also been isolated, including methanogens, sulphate-reducing bacteria and phototrophic purple sulphur bacteria.

Although some halophilic archaebacteria (halobacteria) have minimum NaCl concentrations for growth of $c.15$–20%, many will grow at lower concentrations than previously appreciated and, in some cases, minimum concentrations for growth are as low as 8%. These micro-organisms may be known by the term extreme halophile. The optimum is 20–25%. Four genera are currently recognized, of which *Halobacterium* is most common. All are characterized by pleomorphism and/or irregular cell shapes. In addition gas-vacuolated, square-shaped halophiles have been isolated. Cells are 3–4 μm or larger in size and have not yet been isolated in pure culture. Phages for halophilic archaebacteria have been isolated. These require high levels of Na^+ ions for stability.

Microbial communities in non-alkaline, highly saline environments

Concentrated brines (salt lakes and salterns)

Salt lakes and salterns with NaCl concentrations in excess of 20% are simple environments with short food chains. Many physiological groups, including nitrifying bacteria, sulphate-reducing bacteria and methanogens, are absent at very high NaCl concentrations and the microflora is dominated by halophilic archaebacteria. These are present in numbers up to 10^8 ml^{-1} and the water is often coloured red by the carotenoid pigments. The nature of the dominant microflora and the ecology of the system differs between thalassohaline and athalassohaline environments. There are, however, two major constraints common to each: oxygen deficiency and high light intensity. The solubility of O_2 in concentrated brine is very low at ambient temperatures, concentrations of less than 0.2 mg l^{-1} having been measured in surface waters of the Great Salt Lake. It seems likely, therefore, that halophiles are frequently exposed to stress through O_2 limitation. Many extreme halophiles are gas vacuolated which may be a factor in maintaining the micro-organisms in the upper layers of the brine, although this contention has never been proven. The same role has been proposed for motility and aerotaxis but this again remains unproven. Halophilic archaebacteria do, however, have a number of strategies for growth under conditions of O_2 limitation. Most are capable of growth using NO_3 as the terminal electron acceptor but the environmental significance seems limited since ammonia is the predominant form of nitrogen and a rapid turnover of NO_3 seems unlikely (Oren, 1994).

Many isolates are able to grow anaerobically by the reduction of dimethylsulphoxide to dimethylsulphide, fumarate to succinate and trimethylamine oxide to trimethylamine. The latter mechanism may be of importance in nature since trimethylamine oxide is likely to be present in significant quantities. Halophilic archaebacteria are also capable of fermenting a range of substrates and fermentation may well be an important means of anaerobic growth. Some strains of *Halobacterium* are able to use light energy to drive cellular metabolism when O_2 levels drop too low to support respiration. This mechanism has received considerable attention but is an uncommon feature amongst halobacteria as a whole. Photophosphorylation yields sufficient ATP to prolong viability for extended periods but cannot normally support growth. The mechanism involves synthesis of a modified cell membrane in response to O_2 limitation. This is purple in colour and is often known as the purple membrane. The visual pigment bacteriorhodopsin is responsible for coloration and for the conversion of light to electrochemical energy by pumping protons from the cell interior to the exterior. This creates a cross-membrane potential that the cell can use for other energy-requiring purposes.

The light intensity at the surface of salt lakes and salterns is very high and a general property of halobacteria is the synthesis of carotenoid pigments to act as photoprotectants. *Halobacterium halobium* contains additional retinal-based pigments such as halorhodopsin and has both blue and red light-sensitive systems. The blue-sensitive system mediates the avoidance of blue light, while the red-sensitive system mediates both an attractant response to red light and a much more sensitive avoidance of UV-B irradiation. It is 'tempting to postulate that halophilic archaebacteria use sensory rhodopsin to move into an environment of high light intensity near, but not too near, the surface where exposure to UV-B is potentially lethal. While there the cells rely on the rhodopsin pigments to run several essential ion "pumps". There is, however, little or no evidence to support this scenario.' (Oren, 1994).

Thalassohaline brines

Thalassohaline brines are derived from the evaporative concentration of seawater and all have a remarkably similar ecology. Environments of this type are typified by the solar salterns used commercially for production of salt. A number of salt lakes which are not, or are no longer, in contact with the ocean have a similar ionic composition to salterns and a similar microflora. These include the Great Salt Lake in the US and Lake Assal in Djibouti which is fed by saline salt springs.

During the evaporation of seawater in salterns, sequential precipitation of calcium carbonate and calcium sulphate (gypsum) occurs leaving a remaining hypersaline brine. Precipitation of $CaCO_3$ may involve a moderately halophilic species of *Vibrio*. Sodium chloride subsequently precipitates as halite crystals and dense brines of magnesium chloride remain. These brines – bitterns – are highly hostile to life. The microflora undergoes successional change during concentration. In the early stages the original marine bacteria are replaced by biochemically active, Gram-negative halophiles of relatively low optimum NaCl concentrations for growth. Moderately halophilic bacteria, such as *Halomonas*, increase rapidly in numbers when the NaCl concentration exceeds *c*.7%.

These organisms dominate the microflora at NaCl concentrations up to *c*.17.5%, after which extremely halophilic archaebacteria are of increasing importance. At a concentration of *c*.20% the outcome of competition between halophilic eubacteria and archaebacteria is largely dependent on ambient temperature, with low temperatures favouring eubacteria and high temperatures favouring archaebacteria. The two types are able to coexist in some circumstances. *Haloferax* (285) is the dominant halobacterium at NaCl concentrations of *c*.20%, but is replaced by *Haloarcula* as the salinity increases. *Halobacterium* (286, 287) is dominant at very high NaCl concentrations in the crystallizer ponds, although *Hx. volcanii*, which is highly tolerant of Mg^{2+}, ions may develop after the precipitation of halite. A number of halobacteria produce bacteriocins (halocins), which may enhance their competitive ability. Anaerobic eubacteria have been isolated from anoxic sediments of salt lakes and solar salterns where archaebacteria are unable to grow. Four genera are currently recognized: *Haloanaerobacter*, *Haloanaerobium*, *Halobacteroides* and *Sporohalobacter*, but others may exist.

Bacteria isolated from salt lakes and salterns are heterotrophic and primary production is largely by the eukaryotic alga *Dunaliella*. *Dunaliella* is capable of rapid growth and blooms at cell densities in excess of 10^4 ml^{-1}. The algal blooms are relatively short lived but are often followed by blooms of halophilic archaebacteria whose numbers exceed 10^8 ml^{-1}. Excretion of the osmolyte glycerol by *Dunaliella* and its rapid uptake by halobacteria is thought to be an important factor in archaebacterial blooms (Oren, 1993).

Although both amoeboid and ciliate protozoa have been reported in hypersaline brines, nothing is

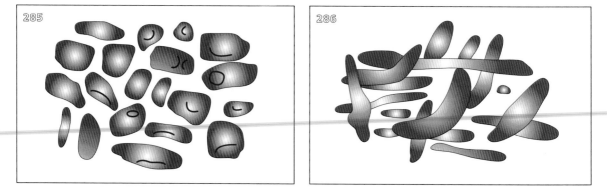

285, **286** Diagrammatic representation of the cellular morphology of extremely halophilic archaebacteria: *Haloferax* (**285**), *Halobacterium* (**286**).

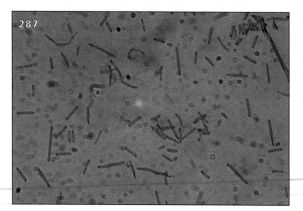

287 Photomicrograph of *Halobacterium*. Like many rod-shaped halophiles, including both extreme and moderate genera, pleomorphism and the presence of irregularly shaped cells is common, especially when grown under sub-optimal conditions. Reproduced by courtesy of Prof. W.D. Grant, Department of Microbiology and Immunology, University of Leicester, UK.

known of their predatory habits (Oren, 1994). In some systems, however, *Dunaliella* is grazed by macroscopic brine shrimps (*Artemia salina*) which can reach a sufficiently high population density to justify commercial harvesting. Brine shrimps and the larvae of brine flies are an important source of organic substrates for heterotrophic micro-organisms. Considerable quantities of chitin may be deposited in sediments, and halophilic, anaerobic, chitin-utilizing bacteria (*Haloanaerobacter*) have been isolated from sediments in salterns in Chula Vista, California (Liaw and Mah, 1992). Bacterio-phages for *Halobacterium* may exist in either the virulent or temperate state, the transition between the two being mediated by NaCl concentration. Phages proliferate only at the lower end of the NaCl growth range when the halobacteria are stressed by the sub-optimal conditions prevailing. At higher NaCl concentrations the phage–host relationship is temperate. The *Halobacterium*–halophage relation-ship is thus well adapted to salterns in that produc-tion of phages is favoured in the less concentrated brine ponds, thus maximizing the chance of phage survival. At concentrations approaching saturation, conditions are optimal for growth of *Halobacterium* and the establishment of the carrier state simul-taneously protects bacteria from extensive phage-induced lysis and provides for the perpetuation of the phage (Oren, 1994). Little is known of the the role of phage in regulating *Halobacterium* populations in nature. Phage numbers increase following dilution of brine by heavy rain but the *Halobacterium* population is destroyed by further dilution. It appears, therefore, that phage develops at the expense of bacteria doomed by other factors without exerting a selective disadvantage on the host.

Stratified microbial mat communities develop in some salterns. The inability of non-specialized micro-organisms to grow at very high NaCl concentrations means that these do not develop in the most concentrated brines but in those with a NaCl concentration of *c*.13–20%. The community structure is generally similar to mats in environ-ments of lower salinity but selective pressures usually mean that fewer micro-organisms are involved. Hypersaline mat communities occur in a number of geographical locations including Salins-de-Giraud in the Camargue region of France, Alicante in Spain, Guerro Negro in Baja California and the Solar Lake in Sinai. The specific composition of the microflora is similar in each case, the purple layer being composed of anoxy-genic phototrophs of the families Rhodospirillaceae and Chromatiaceae. There is variation, however, in the cyanobacterial population forming the green surface layer. In the Guerro Negro ponds and Solar Lake this consists of *Microcoleus chthonoplastes* while *Spirulina* is dominant in the Alicante and *Phormidium* in the Salins-de-Giraud salterns. *Spirulina* and *Phormidium* sometimes coexist with the unicellular cyanobacterium *Aphanothece*. The cyanobacterial population is related directly to the salinity in each environment.

The Salins-de-Giraud salterns are of particular interest in that the mat develops between the sediment and a calcium sulphate (gypsum) crust, forming a physical barrier between mat and over-lying water column. This means that the mat is a closed system, no interchanges taking place with the external environment (288) (Caumette *et al.*, 1994).

288 Structure of microbial mat community. Salins-de-Giraud salterns, Camargue. Structure of zones:
A. *Aphanothece*; B. Gypsum crystals; C. *Phormidium*;
D. Purple bacteria (*Chromatium*; *Thiocapsa*); E. Black sediment containing FeS, gypsum and sulphate-reducing bacteria. After Caumette *et al.* (1994).

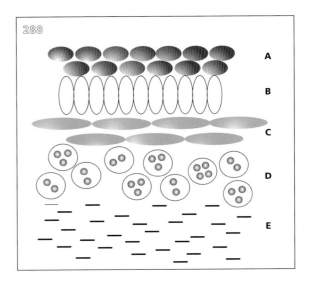

The Salins-de-Giraud mats have a population structure consisting of *Phormidium* (**289**) directly below the gypsum layer, underlain by moderately halophilic members of the Chromatiaceae, resembling *Chromatium* (**290**) and *Thiocapsa* (**291**). This layer is directly above the anoxic sediment, which contains FeS and gypsum and supports a population of the moderately halophilic, sulphate-reducing bacterium *Desulfovibrio halophilus*. Strictly anaerobic, halophilic eubacteria are also present. These organisms are fermentative and of considerable importance in supplying low molecular weight organic compounds and H_2 to sulphate-reducing bacteria. Colourless sulphur bacteria are present in significant numbers throughout the mat and the upper layer of the sediment. In summer, an additional cyanobacterial layer, consisting of yellowish-brown *Aphanothece* develops above the gypsum layer. Sulphate reduction rates in the sediment are high, sulphide produced diffusing upwards. During the day sulphide is reoxidized by oxic and anoxic oxidations, involving chemical oxidation by O_2 produced by cyanobacteria, chemolithotrophic activity of colourless sulphur bacteria and anoxygenic photosynthesis by purple sulphur bacteria. During the dark, the absence of O_2 production by cyanobacteria means that sulphide reoxidation is incomplete and sulphide migrates upward to the gypsum crust. Sulphide in the upper region of the mat is removed during the next illuminated period, it being possible that initial photosynthesis by cyanobacteria is anoxygenic, using sulphide as the electron donor. Longer term seasonal variations occur in that sulphide slowly accumulates during the winter when activity of the photosynthetic microflora is very low. Sulphide levels then fall in spring when photosynthesis is at a very high level, despite increasing rates of sulphate reduction.

Chromatium salexigens and *Tc. halophila* both contain bacteriochlorophyll *a* as photosynthetic pigments but the absorbance maxima differ. This enables the two organisms to coexist, using the same infrared light, but without competition for the same wavelengths (Caumette *et al.*, 1994).

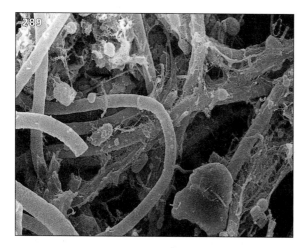

289 The green layer consisting of *Phormidium* of the microbial mats at the Salins-de-Giraud salterns, Camargue. Reproduced with permission from Caumette *et al.*, 1994, © 1994, Federation of European Microbiology Societies.

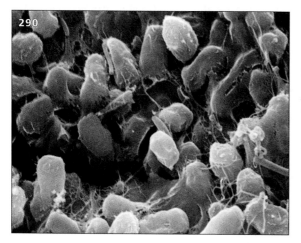

290 Bacteria resembling *Chromatium* in the purple layer of the microbial mats at the Salins-de-Giraud salterns, Camargue. Reproduced with permission from Caumette *et al.*, 1994, © 1994, Federation of European Microbiology Societies.

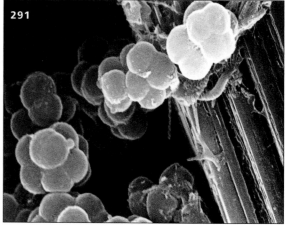

291 Bacteria resembling *Thiocapsa* in the purple layer of the microbial mats at the Salins-de-Giraud salterns, Camargue. Reproduced with permission from Caumette *et al.*, 1994, © 1994, Federation of European Microbiology Societies.

Salterns and salt lakes in tropical and subtropical climates can attain temperatures as high as 41°C (106°F) and ambient temperatures are generally close to the optimum for growth of halobacteria (35–40°C (95–104°F)). Such high temperatures are the exception in temperate climates and the inhabitants of ecosystems such as the Great Salt Lake are usually growing at sub-optimum temperatures. The most striking example of low temperature growth of halophilic archaebacteria is provided by Deep Lake in Antarctica. The highest temperature attained in this lake is 11.5°C (52.7°F) and for much of the year the temperature does not exceed 1°C (33.8°F). This lake is inhabited by *Hb. lacusprofundi* which has a T_{opt} of 31–37°C (87.8–98.6°F) but which is able to grow at *c*.4°C (39°F). It has been estimated that, in its natural environment, this organism undergoes no more than eight doublings per year. Despite this very slow growth rate *Hb. lacusprofundi* is able to completely dominate the microflora due to a lack of competition from other halobacteria, which are unable to grow below 10–12°C (50–54°F).

Athalassohaline brines

Athalassohaline brines, in which ions other than Na$^+$ and Cl$^-$ are predominant, vary in chemical composition. Soda lakes are a special type of highly alkaline athalassohaline brine, supporting a specialized microflora of haloalkaliphiles. *Halobacterium* and other neutrophilic halophiles are unable to grow in these environments.

The best known athalassohaline environment is the Dead Sea. This has been monitored for many years and a wealth of information is available concerning its chemistry and microbiology. The Dead Sea represents an extreme example of an environment where divalent ions contribute the majority of the cation sum and which is highly hostile to microbial life. The normal molar composition is 1.8M Mg^{2+}, 0.4M Ca^{2+}, 1.7M Na$^+$ and 0.14M K$^+$. Halophilic archaebacteria adapted to growth in thalassohaline environments are unable to grow in brine of this composition, primarily due to the toxicity of the Mg^{2+} ion. Archaebacterial inhabitants of the Dead Sea, *Hx. volcanii* and *Hb. sodomense*, are highly adapted to the environment and are not only highly tolerant of Mg^{2+} but have an obligate requirement for the ion. Despite this, the concentration of Mg^{2+} ions is the limiting factor for extensive microbial growth in the Dead Sea. This results from the inhibition of *Dunaliella* at very high Mg^{2+} concentrations and consequent loss of primary production. Development of micro-organisms in the Dead Sea is closely related to the physical structure of the water, especially stratification and the existence of an upper layer of relatively low ion concentration. During normal dry periods the Dead Sea is non-stratified and of a more or less uniform composition with a high concentration of Mg^{2+} and other ions. The water activity level under these conditions is 0.67 or lower. *Dunaliella* is not detectable and archaebacterial numbers fall to an estimated 5×10^5 to 10^6 ml^{-1}. These values are almost certainly overestimates. Significant rainfall leads to stratification with a pynocline at varying depths and a relatively dilute upper layer. This results in a *Dunaliella* bloom, which is short lived, but followed by an archaebacterial bloom when numbers reach in excess of 5×10^7 ml^{-1}. The archaebacterial bloom is longer lived, numbers falling to a stable level of *c*.5×10^6 ml^{-1} until stratification disappears. This usually involves a period of about one year. Extensive growth of micro-organisms in the Dead Sea is thus not a normal phenomenon but occurs as a result of a 'catastrophe' involving the unusual event of abundant rainfall in the catchment area (Oren, 1994).

Solar and mined salt

Sodium chloride precipitates out of solution in the crystallizer ponds of solar salterns as halite crystals. Pockets of concentrated brine are trapped during crystal growth and constitute up to 6% of the crystal weight. Halobacteria are captured in the brine pockets and remain viable for several years, crude solar salt containing 10^5–10^6 bacteria g^{-1}. Survival mechanisms in extreme halophiles have not been elucidated although it has been assumed, without any apparent justification, that parallel processes to those in non-halophiles are involved. A most intriguing development has been the isolation of live halophilic archaebacteria from rock salt buried for more than 200 million years (Norton *et al.*, 1993). Extremely halophilic archaebacteria have been isolated from the interior of intact pieces of rock salt and from sources in the mine, including salt efflorescence on tunnel walls and brine pools. A number of different halobacteria have been isolated including *Hb. saccharovorum* and species of *Haloarcula*. Some isolates are distinct from known genera. The origin of these bacteria remains obscure and it is not known whether or not they represent 'living fossils' that have survived for hundreds of millions of years (Oren, 1994).

Meat curing brines

Meat curing brines have NaCl concentrations in the range of *c*.20–27.5%. Sodium nitrite is present at levels of 0.1–0.15% and, in many cases, sodium nitrate at levels of 0.3–0.5%. The bulk of the population consists of moderately halophilic, Gram-negative rods, phenotypically resembling non-halophilic *Pseudomonas* species. The dominant microflora attains a population density of $10^8–10^9$ ml[1] under normal conditions. Growth can be very rapid but is usually controlled by maintaining the brine at a temperature of less than 5°C (41°F). Growth is also restricted by the presence of sodium nitrite although the bacteria present have a high level of resistance to this salt. Brines are strongly reducing in sub-surface layers and both nitrate and nitrite are used as terminal electron acceptors. The reduction of nitrate is of techno-logical importance in producing the active curing ingredient nitrite. Use of NO_3 and NO_2 as terminal electron acceptors is probably the most important mechanism of anaerobic growth but most of the bacteria present can also use the arginine dihydro-lase pathway for this purpose.

In addition to the main population of moderate halophiles, curing brines contain a sub-population of weakly halophilic bacteria. These are predominantly species of *Vibrio*, such as *V. costicola*, which differ from marine isolates in a number of ways. Such bacteria are unable to grow in the body of the brine. The NaCl content, however, is not homogeneous and relatively low concentrations are found close to the meat where water diffuses out of the tissues.

Hypersaline soils

Hypersaline soils are restricted habitats found in the bottoms of abandoned salterns and other areas previously covered by water of high NaCl content. In many cases there is abundant plant cover provided by xerophilic plants such as *Anthrocnemum* (Quesada *et al.*, 1982). Little information is available concerning hypersaline soils but the microflora appears to consist largely of moderately halophilic and halotolerant bacteria. These have been assigned to a wide range of Gram-positive and Gram-negative genera, the most numerous of which were *Bacillus*, *Alcaligenes* and *Pseudomonas*. Many of the halophilic isolates grew over a very wide NaCl range, suggesting that spatial and temporal variation in NaCl content of the soil favours less specialized halophiles. Small numbers of extremely halophilic bacteria (**292**, **293**), identified with *Halobacterium*, were also isolated, probably from microhabitats of high NaCl concentration (Quesada *et al.*, 1982) Bacteria are present in greater numbers in rhizosphere hypersaline soil than non-root soil and it is probable that the role of bacteria in such soil, and the relationship with plants, is similar to that in non-saline soils.

Leaf surfaces of *Atriplex halimus*

The leaf surfaces of *Atriplex halimus*, a salt-excreting plant, have only recently been recognized as a terrestrial halophilic environment (Simon *et al.*, 1994). Salts and organic matter coat the leaves

292, 293 Bacterial growth on hypersaline soil. Disused salterns in Britanny, France. Although unmanaged, sea water periodically enters some of the salterns and there is considerable, visible, NaCl deposition. The red colouration results from the growth of *Halobacterium* spp., but there is also a diverse population of moderate, Gram-negative halophiles and halotolerant bacteria, predominantly *Bacillus* spp. Illustrations reproduced by courtesy of Liam Varnam.

during the dry season although overnight dew leads to diurnal wetting, followed by repeated desiccation. The leaf surface environment is therefore subject to wide variation in salinity and water activity. The microflora present is of very limited diversity being dominated by an orange-pigmented, halotolerant species of *Pseudomonas*. This has an optimum NaCl concentration for growth of 5% but will grow at concentrations up to 20% and is highly desiccation resistant. Desiccation resistance appears to be related to the protective effect of organic material from the leaf rather than being an intrinsic property of the bacterium.

REFERENCES

Barth, A.H., Cristophi, N., Neal, C. *et al.* (1987) Trace element and microbiological analysis of alkaline ground waters in Oman, Arabian Gulf: A natural analogue for cement pore waters, *Report of the Fluid Processes Research Group of the British Geological Survey*, FLPU 87–2.

Brierley, C.L., Brierley, J.A., Norris, P.R. and Kelly, D.P. (1980) Metal-tolerant micro-organisms of hot, acid environments, *Microbial Growth and Survival in Extremes of Environment* (eds. G.W. Gould and J.E.L. Corry), pp. 39–51, Academic Press, London.

Brock, T.D. (1984) *Thermus, Bergey's Manual of Systematic Bacteriology*, Vol. 1 (eds. N.R. Krieg and J.G. Holt), pp. 333–7, Williams and Wilkins, Baltimore.

Caumette, P., Matheron, R., Raymond, N. and Relexans, J.-C. (1994) Microbial mats in the hypersaline ponds of Mediterranean salterns (Salins-de-Giraud, France), *FEMS Microbiology Ecology*, **13**, 273–86.

Friedman, S.M. (1992) Thermophilic microorganisms, *Encyclopedia of Microbiology*, Vol. 4 (ed. J. Lederberg), pp. 217–29, Academic Press, New York.

Gadkari, D., Schricker, K., Acker, G. (1990) *Streptomyces thermoautotrophicus* sp. nov., a thermophilic CO- and H_2-oxidizing organism, *Applied and Environmental Microbiology*, **56**, 3727–34.

Grant, W.D. (1992) Alkaline environments, *Encyclopedia of Microbiology*, Vol. 2 (ed. J. Lederberg), pp. 73–84, Academic Press, New York.

Haber, R., Stoffen, P., Cheminee, J.C. *et al.* (1990) Hyperthermophilic archaebacteria with the under and open-sea plume of erupting MacDonald Seamount, *Nature*, **345**, 179–82.

Liaw, H.J. and Mah, R.A. (1992) Isolation and characterization of *Haloanaerobacter chitinovorans* gen. nov., sp. nov., a halophilic, anaerobic chitinolytic bacterium from a solar saltern, *Applied and Environmental Microbiology*, **58**, 260–6.

Norton, C.F., McGenity, T.J. and Grant, W.D. (1993) Archaeal halophiles (halobacteria) from two British salt mines, *Journal of General Microbiology*, **139**, 1077–81.

Oren, A. (1993) Availability, uptake and turnover of glycerol in hypersaline environments, *FEMS Micro-biology Ecology*, **12**, 15–23.

Oren, A. (1994) The ecology of extremely halophilic archaea, *FEMS Microbiology Reviews*, **13**, 415–39.

Quesada, E., Ventosa, A., Rodriguez-Valera, F. and Ramos-Cormenzana, A. (1982) Types and properties of some bacteria isolated from hypersaline soils, *Journal of Applied Bacteriology*, **53**, 155–61.

Reysenbach, A.-L. and Deming, J.W. (1991) Effects of hydrostatic pressure on growth of hyperthermophilic Archaebacteria from the Juan de Fuca ridge, *Applied and Environmental Microbiology*, **57**, 1271–4.

Simon, R.D., Abeliovich, A. and Belkin, S. (1994) A novel terrestial halophilic environment: the phylloplane of *Atriplex halimus*, a salt-excreting plant, *FEMS Microbiology Ecology*, **14**, 99–110.

Straub, W.L., Deming, J.W., Somerville, C.S. et al. (1990) Particulate DNA in smoker fluids: evidence for existence of microbial populations in hot hydrothermal systems, 1441–7.

APPENDICES

I Development of the Microflora in a Flooded Gravel Pit

The extraction of gravel, as a constructional material, is an important economic activity in some countries. Pits are drained by mechanical pumping while extraction is continuing, but flooding follows after commercial working ceases. Some are used for recreational use, such as water-skiing, or as marinas, while others have been used as land-fill sites. The gravel pit discussed in this Appendix was, after gravel extraction ceased, allowed to flood. There was also some illegal dumping of land-fill.

The initial microflora developing once the pit was fully flooded, reflected that of the surrounding soil. In this context, it should be noted that the siting of the pit caused very heavy run-off during rain. The initial microflora was dominated by Gram-negative, yellow-pigmented rod-shaped bacteria and 'coryneform' bacteria, many of which were identified as *Arthrobacter* spp. *Bacillus* spp. were also present in relatively large numbers.

As in all environments, the significance of the presence of Bacillus *must be interpreted with caution. The longevity of the endospores means that it is easy to overestimate the importance of the genus in a given ecosystem. This includes foods as well as natural environments. Equally, it can never be assumed that* Bacillus *is present only as surviving endospores.*

Over a period of six months, 'total viable counts' (based on colony counts on quarter-strength nutrient agar medium) fell by a factor of *c.* 10. This occurred over the spring and summer and could not be attributed to low water temperature. Low rainfall, however, meant that run-off from surrounding soil effectively ceased. During this period the qualitative analysis of the microflora showed increasing dominance by Gram-negative rod-shaped bacteria, especially yellow-pigmented. There was also an increase in the number of non-pigmented Gram-negative rod-shaped bacteria. All of the colonies of Gram-negative rod-shaped bacteria were oxidase-positive and presumptively identified with *Pseudomonas* spp.

Over the next six month period (autumn and winter), the 'total viable count' rose slowly to *c.* 10^8 cfu/ml^{-1}. At this stage, the composition of the microflora remained dominated by presumptive *Pseudomonas*. In subsequent months, there was no statistically significant change in 'total viable count', but the diversity of genera increased. Bacteria recovered on quarter-strength nutrient agar medium in numbers in excess of 10^6 cfu/ml^{-1} included oxidase-negative, yellow-pigmented Gram-negative rod-shaped bacteria, purple-pigmented, Gram-negative rod-shaped bacteria (presumptive *Janthinobacterium* spp.), Gram-negative curved bacteria, a number of which produced pink-, or red-pigmented colonies and pale-yellow-pigmented Gram-positive coccal-shaped bacteria, presumptively identified as *Micrococcus* spp. The final sampling, two years after flooding, indicated a stable microflora, although the Gram-positive cocci were no longer detectable.

The presence of high numbers of presumptive Micrococcus *in an aqueous environment is unusual, although the environmental importance of the genus may well be underestimated. The pattern of the development of the microflora, however, illustrates*

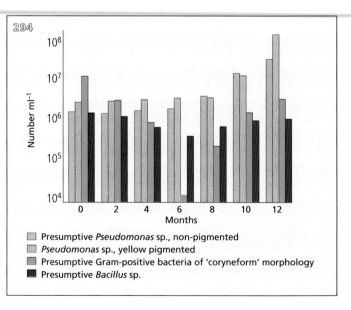

294 Development of the microflora of the water of a flooded gravel pit.

that a 'new' environment, such as the recently flooded gravel pit, can offer transitory niches for a range of micro-organisms, before a more predictable microflora attains full dominance.

The bar chart, **294**, illustrates the change in microbial numbers, recovered on quarter-strength nutrient agar, of the four main 'groups' of bacteria in the water of the flooded gravel pit. In particular, note the overall dominance of the presumptive *Pseudomonas* spp. And consider why there are differences in development pattern between yellow-pigmented strains (population 1) and non-pigmented strains (population 2). The bar chart **295** illustrates the diversity of the population before the disappearance of the Gram-positive coccal shaped bacteria.

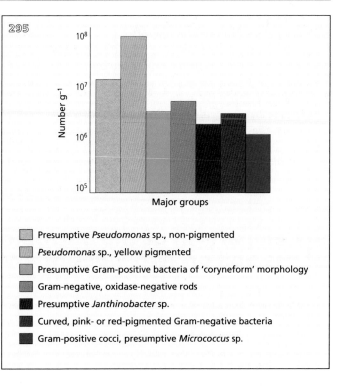

▨	Presumptive *Pseudomonas* sp., non-pigmented
▨	*Pseudomonas* sp., yellow pigmented
▨	Presumptive Gram-positive bacteria of 'coryneform' morphology
▨	Gram-negative, oxidase-negative rods
▨	Presumptive *Janthinobacter* sp.
▨	Curved, pink- or red-pigmented Gram-negative bacteria
▨	Gram-positive cocci, presumptive *Micrococcus* sp.

295 Dominant microflora of gravel pit two years after flooding.

II The Great Goldmine Mystery

As part of a survey of the microbial ecology of a mountain stream – as it made its way from its multiple sources on a mountain to its confluence with a small river in the valley below the mountain – investigation was made of the water issuing from the adit of a disused gold mine. The adit was accessible for a certain distance, the outflow appearing to originate in a closed side-tunnel.

Water samples taken from the stream above the adit outflow had a 'total viable count' (quarter-strength nutrient-agar medium) of 10^7 to 10^8 cfu/ml. This count was higher than expected, but probably reflected drainage from bog pools, the microflora of which reflected that of the underlying mud and was dominated by yellow-pigmented, Gram-negative, rod-shaped bacteria. Despite the obvious presence of peat there was no evidence of acidification.

Water from the mine adit had a 'total viable count' in the order of 10^3 cfu/ml. The microflora consisted of a mixture of white- and yellow-pigmented, Gram-negative, rod-shaped bacteria. There was no visual evidence of slime growth in the water, while the slate walls and remaining wooden structures within the adit were slime-free, although very moist.

A major surprise resulted from analysis of water immediately downstream from the adit outflow. 'Total viable counts' did not exceed 10^3 cfu/ml and there was a total disappearance of yellow-pigmented bacteria, the microflora being dominated by white-pigmented, Gram-negative, rod shaped bacteria. Enzyme profiles obtained by the API-zym™ system suggested that these bacteria shared an identity with those of the adit outflow.

Samples taken *c.* 100 m from the outflow showed an increase in number to 10^5 cfu/ml. Yellow- and white-pigmented colonies were present in approximately equal numbers. The microflora was restored to a 'total viable count' of 10^7 to 10^8 cfu/ml, predominantly yellow-pigmented, Gram-negative, rod-shaped bacteria at the next accessible sampling point; approximately 400 m downstream.

The findings were repeatable and there was no link with seasonal variations in water flow. The area is metalliferous and a number of heavy metals are present. Semi-quantitative analysis, however, showed no evidence of heavy metal pollution. Unusual features were the very sudden decline in the microbial population apparently associated with the adit outflow and the relatively rapid re-establishment of the typical microflora.

In many areas of microbiology, 'hard to explain' findings are a fact of professional life and cause considerable problems in, for example, the food industry. As is probable in this example, the answer often lies in an overall appreciation of the ecosystem, including the physico-chemical background and geobiology, rather than increasingly sophisticated microbiology. The use of genetic techniques, for example, would undoubtedly explore the relationship between the various components of the microflora rather than use the relatively simple tests employed in this exercise. The use of genetic or other sophisticated tests, however, does not solve the central dilemma in examples such as the gold mine, where the key question may well be: 'What have we overlooked?'

III Comparison of the Microbiological Profile of the Soil on Two Sides of a Valley

The north side of a valley is forested with a mixture of coniferous and non-coniferous trees, while the south side is arable land, primarily cultivated for malting barley. Analysis was based on 'total viable counts' together with a simple classification using colony morphology and pigmentation, together with cellular morphology (**296**). The pH (10% w/v in distilled water) was also measured.

Analysis showed that 'total viable counts' were markedly higher on the north, forested side of the valley. This may be attributed to the higher levels of nutrients available and is reflected in differential counts of micro-organisms. The count of microfungi, for example, is much higher in soil from the wooded side of the valley – almost certainly due to the greater level of organic carbon in the soil and the presence of detritus, much of which is highly recalcitrant to degradation by non-filamentous micro-organisms.

The levels of Streptomycetes demonstrate this more vividly. In the south (arable) side of the valley, numbers never exceeded 10^3 cfu g^{-1}, while in the north (wooded) side, numbers were typically in the order of 5×10^5 cfu g^{-1}. Gram-negative, yellow, rod-shaped bacteria were also markedly more prevalent in the wooded soil. In contrast, Gram-positive bacteria of 'coryneform' morpholgy were present in much greater numbers in the arable soil, as were Gram-positive, aerobic endospore-forming bacteria, presumptively identified as *Bacillus* spp. Gram-positive coccal-shaped bacteria, however, were present only in very low numbers (<10^3 cfu g^{-1}) in the arable soil, but in much higher numbers (*c.* 10^4 cfu g^{-1}) in the wooded soils of the north face of the valley. This is an unusual pattern since, during the dry years (when sampling took place) the arable soil became very desiccated, which would normally be expected to favour coccal-shaped bacteria. In contrast, the wooded soil remained relatively moist, conditions which would not be expected to favour coccal-shaped bacteria.

Environmental microbiology, like all other branches of the art and science, produces its conundrums. Another lesson not too late for the learning! Quite simply, it is often not possible to investigate all possible factors affecting microbial growth and survival. It is important to understand, however, that 'What you have got is all that you have got: but it is always more than you think'.

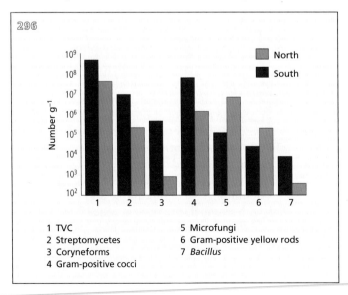

1 TVC
2 Streptomycetes
3 Coryneforms
4 Gram-positive cocci
5 Microfungi
6 Gram-positive yellow rods
7 *Bacillus*

296 Microbiological profile of soil in a valley.

FURTHER READING

Textbooks
General environmental microbiology
Atlas, R.M, Bartha, R. *Microbial Ecology,* 2nd edition. Benjamin Cummings, 1987. *Very general, but a good starting point.*

Hurst, C.J. *Modelling the Environmental Fate of Microorganisms.* American Society for Microbiology, 1991. *The place to begin for would-be modellers.*

Specific principles
Brown, A.D. *Microbial Water Stress Physiology.* John Wiley, 1990. *Underpins knowledge of microbial behaviour in a number of environments.*

Bull, A.T., Slater, J.H. *Microbial Interactions and Communities,* Academic Press, 1982. *Some outdated material, but still worth consulting.*

Henis, Y. *Survival and Dormancy of Microorganisms,* John Wiley, 1987. *Of particular interest because the entire range of microbial life is discussed, not just bacteria.*

Aquatic environments
Pepper, LL., Gerba, C.P., Brasseau, M.L. *Pollution Science.* Academic Press, 1996. *Not dedicated to microbiology, but still a very effective background text.*

Rheinheimer, G. *Microbial Ecology in a Brackish Water Environment.* Springer Verlag, 1997. *Still valuable, but beware outdated material.*

Terrestrial environments
Dommerguez, Y.R, Krupa, S.V. *Interactions Between Non-pathogenic Soil Bacteria and Plants.* Elsevier, 1978. *Fundamental discussions are still appropriate, but care must be taken over detail.*

Paul, E.A., Clark, F.E. *Soil Microbiology and Biochemistry,* Academic Press 1989. *One of the best!*

Richards, B.N. *The Microbiology of Terrestrial Ecosystems,* Longman, 1987. *Comprehensive in coverage, but limited in depth.*

Extreme environments
Da Costa, M.S., Duarte, J.C., Williams, R.A.D. *Microbiology of Extreme Environments and its Potential for Biotechnology.* Elsevier, 1989. *Based on Symposium proceedings – wide-ranging and rigorous in approach.*

Kushner, D.J. *Microbial Life in Extreme Environments.* Academic Press, 1978. *EIderly, but still a worthwhile read.*

Journals and Reviews
Applied and Environmental Microbiology. American Society for Microbiology. *Not entirely dedicated to environmental microbiology, but always a good place to start a literature search.*

FEMS Microbiology, Ecology. Federation of European Microbiology Societies. *Papers are almost invariably of a high standard, many having a strong theoretical base.*

Journal of Applied Microbiology (formerly *Journal of Applied Bacteriology*). Society for Applied Microbiology. *Publishes over a wide spectrum of applied microbiology, including high quality papers concerned with various aspects of environmental microbiology.*

Letters in Applied Microbiology. Society for Applied Microbiology. *The 'little sister' of the* Journal of Applied Microbiology. *Papers are similar in scope and quality, but are usually short communications.*

Environmental Microbiology. Society for Applied Microbiology. *First issue in 1999, but likely to be a valuable information source.*

Many other journals carry papers concerned with environmental microbiology. These include *Microbiology* (formerly *Journal of General Microbiology*), *Canadian Journal of Microbiology* and the *European Journal of Microbiology.*

INDEX

Index

Index